LEARNING AND INNOVATION IN ECONOMIC DEVELOPMENT

NEW HORIZONS IN THE ECONOMICS OF INNOVATION

General Editor: Christopher Freeman, *Emeritus Professor of Science Policy, Science Policy Research Unit, University of Sussex, UK*

Technical innovation is vital to the competitive performance of firms and of nations and for the sustained growth of the world economy. The economics of innovation is an area that has expanded dramatically in recent years and this major series, edited by one of the most distinguished scholars in the field, contributes to the debate and advances in research in this most important area.

The main emphasis is on the development and application of new ideas. The series provides a forum for original research in technology, innovation systems and management, industrial organization, technological collaboration, knowledge and innovation, research and development, evolutionary theory and industrial strategy. International in its approach, the series includes some of the best theoretical and empirical work from both well-established researchers and the new generation of scholars.

Titles in the series include:

Foundations of the Economics of Innovation
Theory, Measurement and Practice
Hariolf Grupp

Industrial Organisation and Innovation
An International Study of the Software Industry
Salvatore Torrisi

The Theory of Innovation
Entrepreneurs, Technology and Strategy
Jon Sundbo

The Emergence and Growth of Biotechnology
Experiences in Industrialised and Developing Countries
Rohini Acharya

Knowledge and Investment
The Sources of Innovation in Industry
Rinaldo Evangelista

Learning and Innovation in Economic Development
Linsu Kim

The Economics of Knowledge Production
Funding and the Structure of University Research
Aldo Geuna

Learning and Innovation in Economic Development

Linsu Kim

Professor of Management, Korea University and Chairman of the Government Reform Council, Korea

NEW HORIZONS IN THE ECONOMICS OF INNOVATION

Edward Elgar

Cheltenham, UK • Northampton, MA, USA

Published by
Edward Elgar Publishing Limited
Glensanda House
Montpellier Parade
Cheltenham
Glos GL50 1UA
UK

Edward Elgar Publishing, Inc.
6 Market Street
Northampton
Massachusetts 01060
USA

A catalogue record for this book
is available from the British Library

Library of Congress Cataloguing in Publication Data

Kim, Linsu.
Learning and innovation in economic development / Linsu Kim.
(New horizons in the economics of innovation)
Includes index.
1. Technological innovations—Economic aspects—Korea (South)
2. Technological innovations—Government policy—Korea (South)
3. Industries—Korea (South) 4. Technology and state—Korea (South)
5. Technology transfer—Korea (South) 6. Industrial policy—Korea
(South) 7. Korea (South)—Economic policy. I. Title. II. Series.
HC470.T4K5422 1999
338.95195—dc21 98–53413
CIP

ISBN 1 84064 026 X

Printed and bound in Great Britain by MPG Books Ltd, Bodmin, Cornwall

Contents

Acknowledgements

The publishers wish to thank the following who have kindly given permission for the use of copyright material.

Carfax Publishing Ltd for articles: 'Innovation Strategy of Local Pharmaceutical Firms in Korea: A Multivariate Analysis', with Youngbae Kim and Jinjoo Lee, *Technological Analysis and Strategic Management*, **1**(1), 1989, 29–44; 'The Dynamics of R&D in Industrial Development: Lessons from the Korean Experience', with Gihong Yi, *Industry and Innovation*, **4**(2), December 1997, 167–82.

Elsevier Science for articles: 'Stages of Development of Industrial Technology in a Developing Country: A Model', *Research Policy*, **9**(3), 1980, 254–77; 'Technology Policy for Industrialization: An Integrative Framework and Korea's Experience', with Carl J. Dahlman, *Research Policy*, **21**, 1992, 437–52; 'Patterns of Technological Change in a Rapidly Developing Country: A Synthesis', with Hosun Lee, *Technovation*, **6**(4), 1987, 261–76.

Greenwood Publishing Group for excerpt: 'Reflections on The Republic of Korea's Acquisition of Technological Capability', with Larry E. Westphal and Carl J. Dahlman in Nathan Rosenberg and Claudio Frischtak (eds), *International Technology Transfer: Concepts, Measures and Comparisons*, Praeger Publishers, 1985, 167–221.

The Institute for Operations Research and Management Sciences for articles: 'Innovation in a Newly Industrializing Country: A Multiple Discriminate Analysis', with Youngbae Kim, *Management Science*, **31**(3), March 1985, 312–22; 'Crisis Construction and Organizational Learning: Capability Building in Catching-up at Hyundai Motor', *Organization Science*, **9**(4), July–August 1998, 506–21.

Macmillan Press for excerpt: 'Absorptive Capacity and Industrial Growth: A Conceptual Framework and Korea's Experience', in Bon-Ho Koo and Dwight H. Perkins (eds), *Social Capability and Long-Term Economic Growth*, St Martin's Press, 1995, 266–87, references.

Oxford University Press Inc. for excerpts: 'Pros and Cons of International Technology Transfer: A Developing Country's View', in Tamir Agmon and Mary Ann von Glinow (eds), *Technology Transfer in International Business*, 1991, 223–39; 'National System of Industrial Innovation: Dynamics of Capability Building in Korea', in Richard R. Nelson (ed.), *National Innovation Systems: A Comparative Analysis*, 1993, 357–83.

The Regents of the University of California for article: 'The Dynamics of Samsung's Technological Learning in Semiconductors', *California Management Review*, **39**(3), Spring 1997, 86–100.

Every effort has been made to trace all the copyright holders but if any have been inadvertently overlooked the publishers will be pleased to make the necessary arrangements at the first opportunity.

Introduction

Economic development through industrialization is a process of acquiring technological capabilities and translating them into competitive products and processes in the course of continuous technological change in the market. The acquisition of technological capability, in turn, is a process of learning, while the translation of technological capabilities into competitive products and processes is a process of innovation. Many studies have shown that more than 50 per cent of long-term economic growth in advanced countries stems from learning and innovation. For this reason, the question often raised is how learning and innovation can be effectively used for economic and social development in developing countries.

Research on learning and innovation is concentrated mainly in advanced countries. Despite the fact that many developing countries have made significant progress in industrial, educational, and technological development, research on learning and innovation in these countries is, however, scanty. Models that capture learning and innovation in developing countries are essential in order to understand the dynamics of industrialization in these countries and also to extend the theories developed in advanced countries.

In advanced countries, innovation through 'learning by research' plays a dominant role in expanding the technological frontier. In developing countries, by contrast, imitative reverse engineering through 'learning by doing' is a common pattern of accumulating technological capabilities. A few newly industrializing economies (NIEs) have, however, made the rapid transition from 'learning by doing' to 'learning by research'. Korea is one of these.

The process of industrial and technological development in Korea is, indeed, an interesting case that offers many useful implications for other developing countries. Korea began as the poorest country among NIEs today with a far lower technical base than other NIEs. As late as 1961, Korea suffered from almost all the difficulties most poor countries are facing today, but it has achieved phenomenal industrial development in the past three decades. Beginning in 1962, the Korean economy grew at an average annual growth rate of almost 8 per cent over the following thirty years. Exports increased from a mere $40 million to $125 billion during the same period. In the mid-1960s, Korea began exporting textiles, apparels, toys, wigs, plywood and other labour-intensive mature products. Ten years later, ships, steel, consumer electronics and construction services from Korea challenged established suppliers from the advanced countries. In the mid-1980s, computers, semiconductor memory chips, video cassette recorders, electronic switching systems, automobiles and other technology-intensive products added to the list of Korea's major export items. In the mid-1990s, Korea is working on the next generation products such as multimedia technology, high-density television, personal communication systems and a new type of nuclear breeder. Such phenomenal growth and structural change in exports may be attributed to many social, economic and technological factors, but most important of all may be the extent of technological learning and innovation in its industry.

Growth in Korea has undergone three stages: duplicative imitation, creative imitation and innovation. Beginning largely with duplicative imitation through the reverse engineering of mature foreign technology, Korean firms moved on to creative imitation, improving and adapting imported technology. Furthermore, a few Korean industries are now generating their own innovation to challenge firms from advanced countries in the international market. Although Korea is experiencing an economic crisis in 1998 due to some macroeconomic mismanagement and insufficient regulation of its financial system, it is a temporary problem, which will be solved, because Korea has a strong technological base and is determined to expand its modern sectors efficiently.

I have conducted in-depth research of more than 200 Korean firms across different industrial sectors including textiles, steel, machinery, electronics, construction, papermaking, cement, shipbuilding, automobiles, chemicals, pharmaceuticals, telecommunications, semiconductors and engineering. On the basis of these studies, I have published over fifty articles in international academic journals on organization design, innovation, strategic management, organizational learning and technology policy. The twelve articles included in this book are selected essays that shed light on some aspects of technological learning and innovation in Korea. Part I covers eight articles on the process of learning and innovation in such industries as consumer electronics, machinery, pharmaceuticals, automobiles and semiconductors. Part II has four articles that discuss issues related to national innovation system and public policy.

My early years

Few academics have as diverse background as I have. Finding mathematics intriguing and relatively easy, I aspired to enter the fields of natural science or engineering. But the Korean War (1950–53) and subsequent chaos interrupted my schooling. Barely finishing high school, I got a job in the government as a junior telecommunications technician. It was during this time that I met Susie Kim, my wife of 32 years happy marriage, at a Christian youth meeting. Christianity has immensely influenced the formation of my philosophy, values and life-style.

Then, I was drafted into the Army, where I spent the following three years (1960–63). During those years, I established a school, first at elementary level and later at secondary level, to educate illiterate children near my military base. Using that as a stepping stone, some of them finished college education later and became professionals. Susie was a fundraiser for my school. She later did something almost impossible in our culture: despite strong opposition from her family, she gave up her postgraduate education with a full fellowship at the University of Rochester in New York to return home to marry me, then a high school graduate with a meagre job. I believed my wife to be a very far-sighted and supportive person, who was always willing to help me grow.

In the following years, I taught myself English and became a chief administrator at a foreign embassy in Seoul. Then, I moved to a broadcasting company to become a news reporter and spent four years there, rising to a radio programme director, an administrator, and eventually a top executive at the age of 31.

After 13 years of working experience, I decided to pursue college education in the evenings (while maintaining my day job) when my wife became a faculty member at

Ewha Women's University in Seoul. Taking advantage of my management experience, I decided to major in business rather than science or engineering. Sadly, the quality of college education, particularly evening courses, in Korea during those years was so rudimentary that I learned little from the undergraduate programme. But it gave me a degree to go further.

Graduate studies at the University of Hawaii and Indiana University

In 1971, I was awarded an East–West Centre grant by the US government to pursue the Master of Business Administration (MBA) programme at the University of Hawaii. It was the first time since elementary school days that I could concentrate purely on study. I really enjoyed studying full-time and the rigorous discipline required in the MBA programme. This changed my career path from management practice to management theory building. I was supposed to go back to the broadcasting company upon obtaining the degree but, during the MBA programme, I came to the important conclusion that management practice to me was a great deal of headache (that is, labour issues, legal issues, endless problems of managing cash flow and so on) but it offers no intellectual challenge. In contrast, academic work was free of headache but full of intellectual challenge. So, I decided to pursue a doctoral programme at Indiana University, which offered me the most attractive scholarship.

In 1973, as a first year doctoral student in management science at Indiana University, I was fortunate to attend a graduate seminar on technology management chaired by Professor James M. Utterback (now at MIT). While I was struggling to grasp technology and innovation issues in the United States, many questions lingered in my mind. How can science and technology, which seem to be the key to development in advanced countries, be effectively used for economic and social development in developing countries? Are theories related to technology management in advanced countries applicable to developing countries? If not, how does technology change in developing countries? And why? I was so inspired by Professor Utterback's seminar that after a year in the doctoral programme I switched my academic specialization from management science (devoted mostly to mathematical modelling and computer analysis) to organization theory (using sociological and strategic management perspectives).

As a late starter in college education, my objective function was to minimize my time in the educational programme. I finished all the course requirements in the new specialization in two semesters. I also completed a comprehensive field examination and defended a dissertation proposal at the end of the second semester. Then, I was named John H. Edwards Fellow, the highest honour given to a graduate student by the University. After only 19 months at Indiana, I left for Boston, where I finished my dissertation in ten months. Although my stay at Indiana University was short, I was fortunate in having a solid training in organization and innovation at the micro level.

My doctoral dissertation, entitled 'Technological Innovation, Environment, Structure, and Policy Implications: Contingency Approaches', attempted to integrate organization theory with innovation. In the course of completing the dissertation, the Sloan Foundation supported me in a visit to Korea to examine the dynamics of technological change in the rapidly changing electronics industry and address several research questions that had lingered in my mind since Professor Utterback's seminar. How does technology change in a developing country and why? Why are some firms more innovative than others in a similar economic environment?

I published three articles out of the dissertation. One of them was 'Stages of Development of Industrial Technology in a Developing Country: A Model' (Chapter 1), which appeared in *Research Policy* in 1980. It presents a theoretical model of how and why Korean electronics firms acquired, assimilated and improved imported mature technologies through the 1970s. It also summarizes how factors external and internal to the firm evolve in a systematic pattern during the evolution of the industry, discusses the applicability of the model to other developing countries and delineates a set of propositions that may be tested empirically. It was later selected as one of the most theoretically contributed articles in the 20-year history of *Research Policy* and its abstract was reprinted once again in the journal in its anniversary issue.

Innovation policy research at MIT and KDI

The purpose of my move to Boston was two-fold. First, I got a research associate position at MIT, where I could enjoy a challenging opportunity to do serious research on technology and innovation. MIT was one of the three institutions (the others being Harvard and Northwestern Universities) that pioneered research and teaching programmes in technology management in the mid-1970s. My research at the Center for Policy Alternatives at MIT in 1975–8 provided me with an academically challenging opportunity to broaden my research experience beyond the micro level and enabled me to integrate micro innovation issues with macro policy issues. Secondly, my wife could pursue her doctoral education at Boston University.

At MIT I was mainly involved in research on an evolutionary theory of product and process innovation in industrially advanced countries, the role of government in enhancing industrial innovation, and the impact of environmental regulations on industrial innovation. I also had a chance to do an extensive field study of the ocean-related industries in the US to examine the impact of a public policy programme (sea-grant programme) on industrial innovation. This project also enabled me to gain a deeper understanding of US industry and society.

Many scholars influenced the formation of my thinking and perspective on technology and public policy during my stay at MIT. These included the late Professor Herbert Hollomon (former Under-secretary of Commerce and President of the University of Oklahoma), Professor James M. Utterback, the late Dr K.N. Rao, Dr Blair McGugan (former Assistant Minister of State for Science and Technology in Canada), and Professor George Heaton (Worcester Polytechnic Institute). Professors Thomas Allen, Edward Roberts and Eric von Hippel at Sloan School of Management also influenced me in approaching technology issues from diverse perspectives. During these three years, I also taught management science and statistics at Boston University.

In 1978, when my wife completed her education at Boston University and became the first Korean to obtain a doctoral degree in nursing, we decided to return to Korea. I joined the Korea Development Institute (KDI) as a senior fellow. KDI is the most prestigious brains trust for the Korean government in the formulation of economic and industrial development programmes. As one of a few non-economists at the institute, I learned a great deal about economists' views and their jargon on the issues of technology and innovation in the process of economic development. My close working association with economists paved the way to my subsequent collaborations with foreign economists. At this time, my wife joined Yonsei University as an associate professor.

My two years at the institute were very productive, allowing me to complete two monographs and many articles. I concentrated mainly on two major projects: science and technology policy in Korea's economic development and a manpower development in Saudi Arabia. The former enabled me to undertake major research on industrial innovation in Korea and led to my first monograph in Korean (co-authored with Professor Jinjoo Lee), *Technology Innovation: Process and Policy*, which was published by KDI Press in 1982. This was the first book on innovation in Korea and was selected by the Korean Management Association as one of the most influential books on economics and management thought in Korea in the 1980s. In the latter, I was invited by the government of Saudi Arabia to conduct major anthropological field research in Saudi Arabia to examine the behaviour of Saudis in their motivation in geographical and occupational mobility. As part of the project, I published another monograph in English, *A Motivation Survey in Saudi Arabia*, published by KDI Press in 1980.

Professorship at KAIST and Korea University

In 1980 I moved to the Korea Advanced Institute of Science and Technology (KAIST), a highly research-oriented postgraduate school of applied sciences, as an associate professor. I teamed up with Professor Jinjoo Lee in developing and teaching the first postgraduate programme on technology management in Korea. The school provided the most favourable research environment. I published many articles on technology, innovation, corporate strategy, and organization design. One of them (co-authored with Youngbae Kim) was 'Innovation in a Newly Industrialising Country: A Multiple Discriminant Analysis' (Chapter 2), which appeared in *Management Science* in 1985. This article identifies different patterns of innovation behaviour associated with four types of the machinery firms in Korea, calling for different approaches to make innovations successful. It also concluded that the majority of important and crucial information for solving technical problems was transferred free of charge from abroad through informal mechanisms rather than formal collaborations with foreign firms.

In 1981, I was invited by the World Bank to participate in a long-term research project (1981–5) on the acquisition of technological capability at the firm level. Other participants, all economists, were Professor Larry Westphal (Swarthmore), Dr Carl Dahlman (World Bank), Professor Alice Amsden (MIT), Professor Sanjaya Lall (Oxford) and Dr Francisco Sercovich (UNIDO). The concept of technological capability was so new to all the participants in the project that it was a great learning experience for all of us. We met frequently in Washington DC for five-day brainstorming sessions, defining concepts, constructing variables and developing a conceptual model of the process of building technological capability in industry. Then we returned home to test them empirically in field studies. The results of the field studies were fed back to revise our model. It was a long process but we were all excited about the importance and significance of our research. This marked a turning point of my research interest from product and process innovation to technological capability building.

My collaboration with these economists was productive not only in this project but also in subsequent projects. I produced many research articles with one or two of them. Two articles are included in this volume. One, co-authored with Westphal and Dahlman, is 'Reflections on the Republic of Korea's Acquisition of Technological Capability' (Chapter 3), which defines and classifies the concept of technological capability. Using the Korean experience as a case in point, it identifies the process of building technological capability in different industries and the relationship between the capability with exports in commodity and technology. The other, co-authored with Dahlman, is 'Technology Policy for Industrialization: An Integrative Framework and Korea's Experience' (Chapter 9), which appeared in *Research Policy* in 1992. This article presents three analytical frameworks that are interrelated into an integrative model. It also examines Korea's science and technology policy as a case in point to demonstrate the usefulness of the model in analysing and evaluating a nation's technology policy in industrialization. It offers many implications for policymakers in developing countries.

In 1985, I moved to a traditional business school by joining Korea University, one of the three major universities in Korea. Although the school was primarily teaching-oriented, it nevertheless provided me with a supportive environment for research and international networking. Most articles in this volume were written at the university. Co-authored with Youngbae Kim and Jinjoo Lee, 'Innovation Strategy of Local Pharmaceutical Firms in Korea: A Multivariate Analysis' (Chapter 4), which appeared in *Technology Analysis and Strategic Management* in 1989, shows how and why local pharmaceutical firms with different levels of technological capability and scale behave differently in innovation strategy. Chapter 5 (co-authored with Hosun Lee), 'Patterns of Technological Change in a Rapidly Developing Country: A Synthesis', which appeared in *Technovation* in 1987, presents a general model of technological change by synthesizing my previous works through the mid-1980s. The model shows that industries using different operations technology (small batch or unit production, large batch or mass production, or continuous process production) show different strategy in innovation. The scale of operations at the firm level explains differences among firms within the same industry.

For developing countries, technology transfer from advanced countries is a major source of new knowledge. This subject is covered in Chapter 6, 'The Pros and Cons of International Technology Transfer: A Developing Country View', which appeared in a book published by Oxford University Press in 1991. This article analyses various modes of international technology transfer and discusses the pros and cons associated therewith. It also offers many implications for both suppliers and recipients who want to maximize the benefits from international technology transfer.

During this period I was also involved in management consulting of leading Korean conglomerates including LG and Samsung. My commitment as a resident consultant with several subsidiaries in Samsung continues even today. In addition, I developed analytical frameworks and methodologies for research on corporate culture. This led to opportunities for me to consult many Korean enterprises, diagnosing and prescribing for them to bring about management innovations, which are important for making organizations more innovative technologically.

The World Bank project in the early 1980s and subsequent research on technological capability helped me build an important intellectual foundation to take part in another long-term international project (1986-91). This exploratory project pulled together leading scholars in innovation from 15 countries to produce a book on national innovation systems in a comparative term. The participants included, among others, Professors Richard Nelson and Hugh Patrick (Columbia), Carl Dahlman (World Bank), Giovanni Dosi (Rome), Charles Edquist (Linkoping), Beng-Ake Lundvall (Aalborg), Jorge Katz (ECLAC), David Teece and David Mowery (Berkeley), Keith Pavitt and Chris Freeman (Sussex), Nathan Rosenberg (Stanford), Jon Sigurdson (Lund), Luc Soete (Maastricht), Morris Tuebal (Jerusalem) and myself. Our basic strategy was to meet together several times so that all of us could influence one another in drawing a general consensus on the concept of national innovation systems and discuss chapter drafts to help each other towards something that was coherent. Columbia, Maastricht, Sussex and Stanford Universities hosted these meetings. The project was carefully designed to illuminate the institutions and mechanisms supporting technical innovation in the various countries, the similarities and differences across countries and how these came to be. This helped determine the extent and manner in which these differences explained variation in national economic performance. It was a fascinating and rewarding intellectual venture for us to learn much from each other. This project enabled me to broaden my perspective even further in looking at technology issues from a national system perspective. The collection of outputs, *National Innovation Systems: A Comparative Analysis* (editor: Richard Nelson), was published by Oxford University Press in 1993. Chapter 10 in this volume, 'National System of Industrial Innovation: Dynamics of Capability Building in Korea', is a chapter contributed by me in the edited volume. This article evaluates Korea's national innovation system in the past to answer how Korea achieved a phenomenal industrial growth in the 1960s and 1970s. It also analyses the strengths and weaknesses of Korea's contemporary national system of industrial innovation in the late 1980s and discusses problems in sustaining growth in the future. Some of the problems discussed in this article are proven to be underlying factors that have led to the economic crisis at the end of 1997.

Because each of the articles published in international academic journals could cover only one or two narrowly focused subjects, a monograph was needed to synthesize these diverse issues. Then in 1993, Columbia University invited me to teach a year as Visiting Professor at its Business School and to conduct research at its East Asian Institute. This opportunity gave me an intellectually exciting environment which enabled me to begin writing the monograph to synthesize my previous works, to integrate them with more recent research findings and to test some of my ideas with my graduate students at Columbia. In this process, many helped me clarify my thinking on the complex process of technological learning at the organizational level. These included Professors Richard Nelson, Hugh Patrick, Ming-Jer Chen and Michael Tushman.

In 1994, the Institute of New Technology of the United Nations University (UNU/INTECH) invited me to spend the summer in Maastricht, the Netherlands to give seminars and advise PhD interns on their research projects. This gave me the opportunity to undertake an extensive literature search and to strengthen the theoretical underpinnings of the monograph I was writing. Professor Charles Cooper, Director, and Shulin Gu were sources of inspiration. The outcome of the work, *Imitation to Innovation: The Dynamics of Korea's Technological Learning*, was published by Harvard Business School Press in 1997.

Policymaking in the Government

In 1996–8, I was seconded to the Korean government to serve as President of the Science and Technology Policy Institute (STEPI), a think tank and R&D funding organization. This post afforded me numerous opportunities to apply my thinking into policymaking, to get deeply involved in decision making related to national policy formulation, and to have a first-hand understanding of the strengths and weaknesses of the public sector in Korea. Despite a hectic schedule in managing STEPI and getting involved in policymaking and institutional networking with the government agencies, I continued to do academic work, conducting research and writing journal articles.

During this period, my meeting with Professor Ikujiro Nonaka of Hitotsubashi University in Tokyo and his recent book with Professor Hirotaka Takeuchi, *The Knowledge-Creating Company: How Japanese Companies Create the Dynamics of Innovation* (New York: Oxford University Press, 1995), influenced me greatly in focusing more on the process of learning and knowledge creation at the organizational level. Two of the articles I wrote on learning are included in this book as Chapters 7 and 8. While articles mentioned earlier discuss learning and innovation issues in the duplicative and creative imitation stages of industrialization, articles hereafter address not only issues related to the first two stages but also those related to the innovation stage in Korea. 'The Dynamics of Samsung's Technological Learning in Semiconductors', Chapter 7, which appeared in *California Management Review* in 1997, explains the process of how Samsung, as an imitator, has expedited its technological learning to leapfrog over established competitors to become an innovator and the leading memory chip producer in a decade. Chapter 8, 'Crisis Construction and Organizational Learning: Capability Building in Catching-up at Hyundai Motor', which appeared in *Organization Science* in 1998, presents a theoretical framework on organizational learning. It then discusses Hyundai Motor as a case in point to illustrate how constructed crises may be used as a strategic means to expedite learning.

Chapter 11 (co-authored with Gihong Yi), 'The Dynamics of R&D in Industrial Development: Lessons from the Korean Experience', which appeared in *Industry and Innovation* in 1997, was also written during my tenure at STEPI. This article addresses issues related to the interactive evolution of R&D activities in universities, government research institutes, and industries in Korea and discusses their implications for other developing countries. Finally, 'Absorptive Capacity and Industrial Growth: A Conceptual Framework and Korea's Experience' Chapter 12, which appeared in a book published by St Martin's in 1995, introduces the concept of absorptive capacity as an analytical framework and discusses issues related to Korea's past progress, current problems, and future prospects. It shows how a developing country could lose its international competitiveness when it fails to keep building its absorptive capacity.

In 1998, the government appointed me the Chairman of Government Reform Council in the face of a major economic crisis in Korea. The government thought my experience in the public sector as the President of STEPI, my expertise in innovation, and my previous experience as a top executive and a management consultant in the private sector would be a useful background to introduce corporate management concepts into the public sector organizations. This Council is responsible for reforming the public sector to be small and efficient. For this purpose, it deliberates issues related to major restructuring and management innovation of central and local government administration, state-owned enterprises, government research institutes and quasi-government organizations. These tasks are formidable but imperative in the face of a major economic crisis in Korea.

By now, readers see that the twelve articles included in this volume are products of a long process of learning and innovation on my part. On learning, I owe a great intellectual debt to the aforementioned eminent scholars in advanced countries. On innovation, I have struggled a great deal to develop new theories of learning and innovation in developing countries. In this process I have come to realize once again the importance of a basic theory-building principle that theory-generation should precede theory-testing. Theory-generation needs to take an inductive approach, carefully identifying many relevant variables in a broad perspective and their relational patterns in a new, ill-structured context. In contrast, theory-testing may take a deductive approach, generating and statistically testing a set of hypotheses on a limited number of variables from existing theories in a relatively well-structured context. Both theory-generation and theory-testing play important, complementary roles in theory-building, but what is needed for theory-building in developing countries at this point in time is theory-generation research. All articles reproduced in this book take that approach.

PART I

LEARNING AND INNOVATION IN INDUSTRIES

[1]

SUMMARY

This article presents a model of development stages that shows when and why local firms in a developing country acquire foreign technologies, how they assimilate and improve imported technologies to strengthen their competitiveness, and what and when external influences affect the process of technological change in these firms. Industrial technology in Korea's electronics industry has developed through three salient stages. Initially established through the implementation of imported foreign technology, local firms in the industry then accumulated experience in product design and production operation which provided a basis for limited indigenous efforts for the assimilation of imported technology. Finally, increased market competition in local and international markets and increasing capability of local personnel, together with assimilation of foreign technology, led to gradual improvement of foreign technology. This pattern is also evident in the history of manufacturing industries not only in Korea but also in other countries. The article also presents several propositions for future research.

Research Policy 9 (1980) 254–277 *North-Holland*

stages of development of industrial technology in a developing country: a model

by
Linsu KIM*
Korea Development Institute, Seoul, Korea

INTRODUCTION

This article presents a model of three development stages that shows how industrial technology develops over time and how the characteristics of international transfer of technology, the production process structure, and local firms' innovation strategies evolve in response to the changing competitive environment in an industrializing coutry, Korea.

Technological change is known to have been a major determinant of national development. In industrialized economies, many studies have shown that more than 50 percent of long term economic growth stems from technological changes which improve productivity or lead to new products, processes or industries [7]. For this reason, the question often raised is how can science and technology, which appear to be the key to development in the more advanced countries, be effectively used for economic and social development in the less developed regions of the world.

Previous studies have sought to answer many important questions related to science and technology and their role in the development of less developed countries. These studies may be roughly classified into two categories. Studies in the first category are on science and technology and their role in the development of LDCs. They are mainly concerned with the development and role of the local scientific community, the role of education, and effects of science and technology policy on industrial development. Moravcsik [17] lists almost 500 publications on various aspects of this issue. Studies in the second

* The author undertook this research when he was at the Center for Policy Alternatives at the Massachusetts Institute of Technology. He is indebted to his colleagues, J. M. Utterback, S.K. Fung, T. Ozawa, and K. N. Rao at MIT for their critical comments on an earlier draft. The research underlying this article was supported by the John H. Edwards Fellowship of Indiana University and by a grant from the Alfred P. Sloan Foundation to the Center for Policy Alternatives at MIT.

category are concerned with the international transfer of technology to LDCs; that is, the role of multinational firms in technology transfer, the problems associated with importation and adaptation of foreign technology, and the role of local R&D institutions. Steele [24] lists over 200 studies in this area.

This study is a distinct departure from the previous studies, in that it reverses the approach, looking at the level of firms to identify the pattern of when and why local firms acquire foreign technologies, how they assimilate and improve imported technology to strengthen their competitiveness in a changing environment, and when government programs and other external factors affect the process of technological change in firms. This approach is useful and important, since government uses a variety of programs within the general framework of their industrial policies in order to influence the process of technological change at the firm level. Similarly, both multinational companies that supply foreign technologies and local R&D institutions which help local firms import, adapt and improve such technologies also interact primarily at the level of firms.

This article first presents an three-stage model of when and how local firms in a LDC import and implement, assimilate, and improve foreign technologies in response to the changing competitive environment. This model is based on the primary data collected from 31 consumer and industrial electronic products manufacturers and the secondary data on the electronics industry in Korea. The applicability of the model to other industrial sectors in Korea and other countries is then discussed. Finally, the article offers a set of propositions that can be researched in future in other industries in other developing countries.

Korea is one of the most rapidly industrializing countries in the world. Its GDP increased at the annual rate of over 10% between 1965 and 1975, one of the highest growth rates. Its manufacturing sector has grown during the same period with the annual rate of 22.4%, noticeably faster than the economic growth in general [26].

The electronics industry in Korea in which this study was undertaken is the fastest growing industry in Korea. Since 1966, the industry production volume has grown more than 50% every year. Exports have increased more than 100 times from 1966 to 1975 and the number of firms has increased from 70 to 457 during the same period. Of these 457, 102 firms produced the end products in 1975 [6]. The 31 firms were selected from these 102 firms on the basis of 'chunk' sampling techniques so as to have representative samplings on an *a priori* basis through multi-stages (see [14] for the sampling technique). Firms covered in this study produce TV sets, radio, stereo sets, tape recorders, communication and broadcasting equipment, calculators, and simple testing instruments. These products accounted for more than 85% of the total end product production in the electronics industry in 1975.

STAGES OF DEVELOPMENT OF INDUSTRIAL TECHNOLOGY: A MODEL

This study does not include wholly-owned foreign subsidiaries, as they, particularly off-shore assembly firms, behave differently from national and joint-venture firms.*

In recent decades, developing countries have employed various efforts at industrialization. The NAS's ad hoc panel of the Board on Science and Technology describes the general pattern of development for these countries in the following terms:

> The principal industrialization strategy of most LDCs during the early postwar years is said to have been import substitution; domestic production, under the shelter of a high tariff wall and usually under monopoly conditions, of goods that were formerly imported.
>
> By the end of the 1960s, export promotion as a strategy for development had become a full partner of, if not a replacement for, import substitution. This change increased LDC interest in strengthening the technological capabilities of local enterprise – exporters must sell at competitive prices. A basic reason for fostering indigenous RD&E is to combine engineering skills, which are becoming more plentiful in the LDCs, with first hand knowledge of local productive capabilities and social customs and both local and foreign product requirements [19].

Like many other developing countries, Korea has employed this pattern of industrialization [25]. The electronics industry was initially established in response to a market opportunity created and protected by a government im-

* Manufacturing subsidiaries of foreign corporations have a high dependence on their parent companies for product design and production techniques. For example, off-shore assembly subsidiaries perform no more than assembly operation of foreign components and parts for reexport purpose. In this case, products and their components and production processes are developed by parent companies. Even foreign subsidiaries that manufacture products for local markets perform no more than adaptive engineering of product design and production techniques developed by parent companies to local conditions as viewed necessary by the latter. In contrast, local firms evolve in their technological capabilities from implementation, to assimilation, and to improvement of foreign technologies [1, 2, 10].

port substitution policy. In most cases, product design and production technologies had to be imported from overseas. Production at this stage was merely an assembly operation of foreign components and parts, and the firm's technological concern was essentially for the *implementation* of a production operation. There were few indigenous efforts to undertake technological change in products and processes due to the lack of both technological capability and market incentives to bring about change.

As product design and production techniques transferred from overseas were successfully implemented at several firms, technical knowhow was quickly diffused within the country, resulting in an increased number of firms competing in the same market. Accumulated experience in product design and production provided a basis for limited indigenous efforts for the *assimilation* of foreign technologies.

Finally, assimilation of foreign technology and increased market competition together with increasing capability of local scientific and engineering personnel led to the gradual *improvement* of foreign technology. Product improvement and cost reduction became an important basis for competition not only in the domestic market but also in international markets which became increasingly important. Indigenous efforts for technological change became important to improve products and to increase productivity, leading to a relatively more efficient, integrated and larger production system.

In short, there seem to be three salient stages in the development of industrial technology in the electronics industry in Korea: Implementation, Assimilation, and Improvement, named for the predominant technological task of each stage. These three stages will be examined in more detail.

IMPLEMENTATION

Though the AM radio set has been assembled in small scale 'garage-operations' in Korea since 1958, the electronics industry in Korea can be said to have begun in the mid-1960s, with the production of black and white TV sets, stereos and radio communication equipment through the international transfer of production technology [10]. When the government banned the imports of consumer electronic products as a means of import substitution, this measure plus control of foreign investment and contraband goods in the black market provided an attractive business opportunity for local entrepreneurs in the protected local market which had previously been dominated by foreign products. However, there was no local capability to establish production operations. Entrepreneurs who had previous trade contacts with foreign firms quickly imported 'packaged' technology from overseas which included assem-

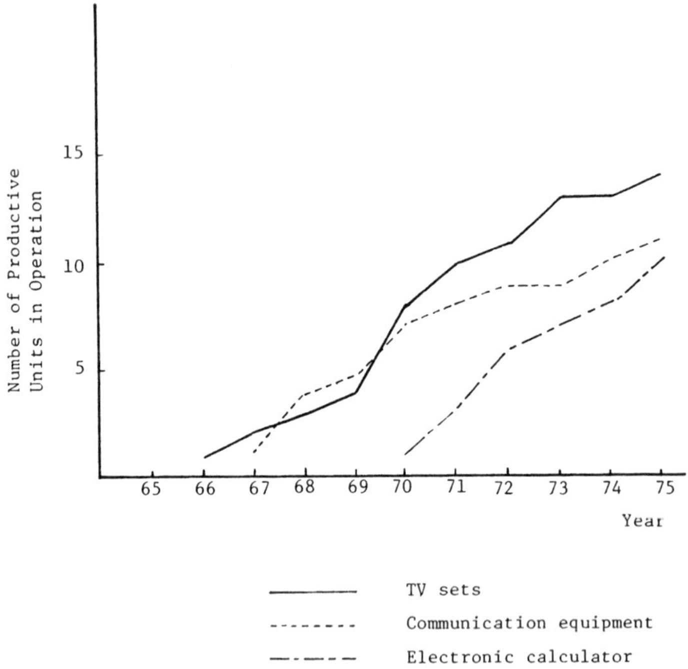

Fig. 1. Number of productive units in operation, 1965–1975.

bly processes, together with product specifications, production knowhow, technical personnel and component parts.

The import of such a 'packaged' technology led to the emergence of new productive units and the electronics industry in Korea. The productive unit refers to the simple firm or electronics division within a large firm. Fig. 1 depicts the growth of the industry in terms of the number of productive units in operation in selected product lines. Of 14 television manufacturers in operation as of 1975, the first four were established through the transfer of 'packaged' technology from Japan and Holland. A similar pattern was also evident, as shown in fig. 1, in communication equipment, electronic calculators, and other electronic products. International transfer of 'packaged' technology provided an initial base for the technological development of the electronics industry in Korea.

Production in this stage was merely an assembly operation of foreign components and parts with equipment purchased from overseas. In the case of black and white TV sets, all parts except the case, resistors, capacitors, etc. were imported from abroad in the first year of production. This was also true in the case of other electronic products introduced during the same period. Simple parts were quickly produced locally as the finished products created the market for the parts industry; however, more complicated parts were imported from overseas.

The production process pattern was inefficient because of low labor costs and little cost pressure in the protected market. No data are available for the 1960s, but low labor cost is evident in a study conducted in 1970 which found that labor costs accounted for only 5.1% of total costs of a 19" black and white TV set, which requires a relatively labor intensive assembly operation [9]. This implies low labor costs relative to other input factors, possibly suggesting that low labor productivity may not be problematic to manufacturers in this stage. The analysis of data collected in this study from 31 productive units shows that operations technology rigidity of Hickson, et al. [8] * is positively correlated ($r = 0.36, p < 0.05$) with indigenous technological capability.** The operations technology rigidity refers to the degree of rigidity within the production process whereby inputs are converted into outputs. A high degree of rigidity indicates a system-like, efficient process, while a low degree of rigidity indicates a loosely organized, inefficient process. The indigenous technological capability refers to the degree of indigenous capability of local firms in developing new products and is significantly related to the age of the productive units under study ($r = 0.50, p < 0.001$), indicating that those at the early stage have low indigenous technological capability and those at the subsequent stages have high indigenous technological capability. Thus, the indigenous technological capability was used in this study as a surrogate to measure technological development stages. To recapitulate, the above relationship indicates that productive units at this stage tend to have more loosely organized, inefficient production processes than those in the subsequent stages. At the industry level, a study shows that a productivity per

* Hickson, et al. [8] developed an 8-item scale to indicate the degree of rigidity. This scale measures workflow continuity, specialization of equipment, variability of operational sequence, necessity of buffer stock, possibility of rerouting work, and interdependence among subprocesses.

** The indigenous technological capability is measured by a 6-point scale. A productive unit which performs a simple assembly operation of 100% foreign parts for foreign-designed products is score 1, while a productive unit that develops and produces unique products, which are not an imitation of foreign products is scored 6. In the case of multiproduct firms, a major product line was used for this measure as well as operations technology rigidity.

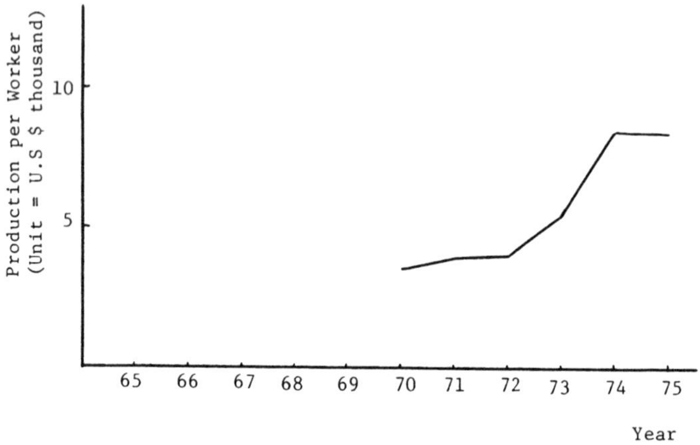

Fig. 2. Labor productivity of the electronics industry in Korea, 1968–1975. Source: Table 23 in [11].

worker per year was US$3,600 in 1970 compared to US$8,340 in 1975, as shown in fig. 2 [11.]

During this initial stage, the immediate technological task was the *implementation* of the transferred foreign technology in order to produce products whose technology and market had been tested and proven elsewhere. For this purpose, only limited engineering (E) efforts were required.

Many productive units interviewed in this study indicated that the technical assistance from foreign experts dispensed to recipient firms was of significant help in debugging problems in the initial implementation of production operations. Of eight productive units under study that employed foreign experts during the early period, seven of them indicated that technical assistance was useful but their utility diminished rapidly within one year as the local technicians acquired production and product design experience.

What are the external stimuli that influence the pattern described above? To answer this question, top managements of the 31 productive units under study were asked to indicate the degree of influence of each subcomponent of task environment on technological change in products and processes. The task environment refers to individuals, organizations, and other social systems external to the firm that affect directly the formulation and attainment of organizational objectives. This study modified Duncan's typology [3] of the task environment and delineated 5 components with 20 subcomponents, as shown in table 1, so as to make them more appropriate in the Korean con-

Table 1
Correlation[a)] between external environment and indigenous technological capability (N=31)

External environment[b)] Components	Subcomponents	Indigenous technological capability[c)]
Customer Component	Domestic distributor	0.70[d)]
	Overseas distributor	
	Domestic consumer	0.56[d)]
	Overseas consumer	−0.31[e)]
Supplier component	Raw material supplier	−0.25[f)]
	Equipment supplier	−0.53[d)]
	Technology supplier	
	Overseas capital supplier	
	Domestic fund supplier	0.25[f)]
Competitor component	Overseas competitor	0.32[e)]
	Local competitor	0.79[d)]
	Competitor for supplier	0.51[e)]
Government component	Government's control on industry	
	Government's incentives for innovation	−0.50[e)]
	Import and export policy	−0.35[e)]
	Control over import of foreign technology	
Technological component	Technological requirement of the industry	
	Development of new technology	
	Availability of local research institutes	
	Availability of local technical information center	

a) Pearson's product moment correlation coefficients.
b) Top managements were asked through interviews to indicate the degree of influence each subcomponent had on technological change in products and processes. Responses are scored from 5 (very strong influence) to 1 (no influence at all).
c) Indigenous technological capability is measured by a 6-point Likert Scale. A productive unit that performs a simple assembly operation of 100% foreign parts for foreign designed products is scored 1, while a productive unit that develops and produces unique products, which are not imitation of foreign products is scored 6. A blank indicates a statistically insignificant relationship.
d) Significance level < 0.001.
e) Significance level < 0.05.
f) Significance level < 0.1.

text.* The degree of influence on technological change was then related to the degree of indigenous technological capability in an attempt to identify how a set of external influences change over different stages of industrial technology development. Thus, an inverse correlation, as shown in table 1, indicates environmental subcomponents that are perceived influential in technological change by productive units at the early stage of technological development, while a positive correlation indicates those subcomponents that are perceived influential by productive units at the later stage of development. Government and multinational firms were important sources of external influence on technological development in this stage. Table 1 shows that indigenous technological capability is inversely related to import and export policy, government incentives for technological change, raw material suppliers, equipment suppliers, and technology suppliers. It indicates that these environmental subcomponents wer perceived as significantly affecting technological change in products and processes by productive units with a low degree of indigenous technological capability at the implementation stage. In other words, government policy was a major influence in this stage in two ways: first, the transfer of foreign technology establishing the industry occured when the government implemented import substitution and market protection policies and second, various government incentive programs were also perceived as important stimuli for technological change by industry. And multinational firms played an important role in establishing the industry as suppliers of technology, technical personnel, capital, equipment and component parts.

ASSIMILATION

Once product design and production technologies transferred from overseas were successfully implemented at several productive units, such technologies were quickly diffused within the country. Fig. 1 shows that by 1975, 14 productive units were in the black and white TV market, 11 in the radio communication equipment market, and 10 in the electronic calculator market. However, those productive units that entered the market later were not established through the international transfer of 'packaged' technology, but

* Duncan [3], in his study of the external environment for manufacturing and R&D organizations in the U.S., suggests, as the task environment, five components: customers, suppliers, competitors, socio-political, and technological. Each component has several subcomponents with the total of 13 altogether. The rationale for 20 subcomponents in this study can be found in [12].

through the transfer of local entrepreneurial and technical personnel from existing productive units. 10 out of the 14 TV producers, six out of 11 radio communication equipment producers and 6 out of 10 electronic calculator makers are in this category. The mobility of experienced technical personnel was evidently the major mechanism for the diffusion of production technology within the country.

When several productive units (established either through the international transfer of technology or through its diffusion within the country) produced a similar product within the same market, the result was noticeable competition among them. Under such conditions each productive unit diversified its product to satisfy the needs of the different segments of the market on one hand and to obtain a bigger share of the market on the other hand. For instance, productive units that began TV production in the mid-1960s entered the market with one or two different 19″ models. By 1970, five producers carried 18 different kinds of 19″ sets in addition to 9″, 16″, 17″ and 23″ models [10]. Product diversification became the prevailing marketing strategy during this stage.

This pattern of competitive strategy required indigenous efforts for the *assimilation* of foreign technologies in order to produce differentiated products. At the same time, competition, not only in the domestic market but also in the international market which was becoming increasingly important, led to more pressure to reduce costs by the manufacturers.

While the foreign experts working in the recipient productive units played an important role in the short-term implementation of imported technology in the first stage, local technical personnel trained at the technology supplier firms played an important role in the assimilation of foreign technology, and product and process development. Technical emphasis in this stage was placed upon engineering and limited development (E&D), rather than research. The finished product industry created a local market for the parts industries. Local components and parts manufacturing units emerged within or outside the finished product manufacturers in response to the local demand, many through the international transfer of production technology. Resistors, capacitors and speakers were locally manufactured soon after AM radio sets were first assembled locally. TV picture tubes and tuners were first made in Korea through the international transfer of production technology four years after black and white TV sets were first assembled. As shown in fig. 3, the part substitution rate for TV sets rose from less than 30% in 1966 to 50% by 1969 and to over 90% by 1975. A similar pattern was also observed, as shown in fig.3, in electronic calculators, radio communication equipment and other products.

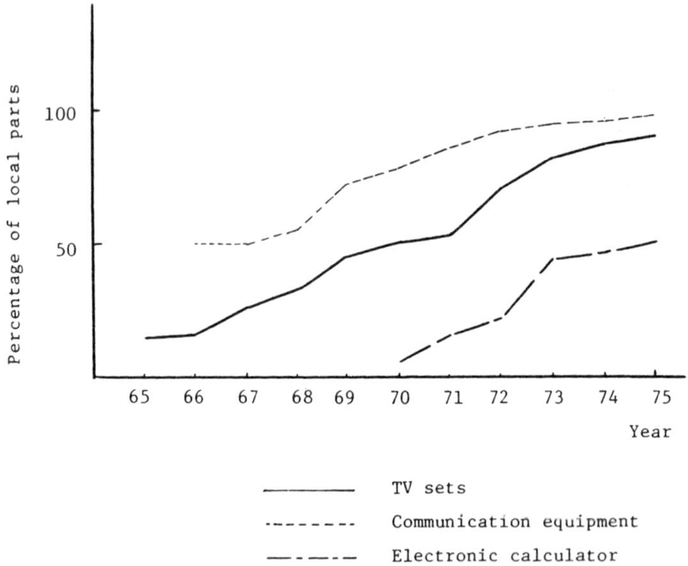

Fig. 3. Rate of parts substitution, 1965–1975.

Like the finished product industry, the rapid development of the local part industry can be largely attributed to the government's import substitution policy. The government encouraged local firms to develop a supplier base by providing credit at preferential interest rates. More importantly, imports were only allowed if the user provided evidence of inferior or insufficient local supply. These measures have prompted finished product producers to integrate backwardly or to assist local suppliers to assure themselves of a reliable source of supply at competitive prices. In short, the emergence of the local parts industry was a characteristic of the Assimilation stage of the finished product. The growth of the electronics industry also provided a stimulus for the development of industrial machinery and machine tool industries. Equipment and machine tools previously imported were gradually substituted by local products.

On the basis of local market experience and in response to stimuli from OEMs (original equipment manufacturers) and other overseas buyers, and to incentives of the government export promotion policy, many productive units developed export markets. The government played a major role in the export expansion. The Electronics Industry Promotion Law of 1969 was aimed at developing the industry as an export oriented industry. Long term loan at low interest rates, duty free for production machinery, accelerated depreciation,

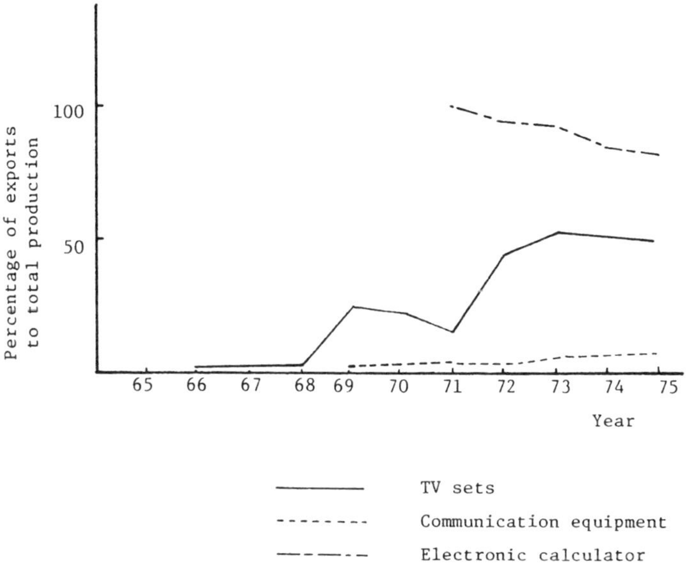

Fig. 4. Percentage of exports to total production, 1965–1975.

and overseas marketing research service were among government programs that were designed to facilitate export expansion. That is, export promotion was gradually emphasized not only as an individual productive unit's growth strategy but also as a national policy. Fig. 4 shows the increase in the percentage of exports to the total production of selected electronic products between 1965 and 1975. For example, while TV production increased from 10,000 sets in 1966 to over 1 million sets in 1975, the percentage of exports to the total TV production rose from 2% in 1968 to over 50% in 1975 (5,11).

But not all exports grew from production to satisfy local demands. Some new products were introduced in this stage in response to the overseas market demand before a sizeable local market was created. For example, previous export experience and trade contacts established earlier led to the production of electronic calculators in Korea solely in response to the growing overseas market demand. The local market for this product developed slowly as seen in fig. 4. Tape recorder and electronic digital watch production also showed a similar pattern. In summary, once the electronics industry was established and general production technology was widely diffused and accumulated, export markets became not only a source of stimulus for further development of the industry in terms of size and technology, but also a source of stimulus for the development of new product lines solely for overseas markets.

What are the external sources that influence the pattern in this stage? As imported technologies were diffused and competition built up in the local as well as the international market, sources of external influences for technological change in products and processes gradually shifted from those of government and multinational firms to those of consumers, distributors, and competitors. Table 1 shows that indigenous technological capability is positively related to domestic distributors, domestic consumers, domestic fund suppliers, local and overseas competitors, and competitors for supply. It indicates that these environmental subcomponents were perceived as significantly affecting technological change in products and processes by productive units with a high degree of indigenous technological capability. In other words, changing needs of domestic consumers, custom orders of overseas distributors and original equipment manufacturers became important sources of stimulus for new product design and development at this and subsequent stages. New products introduced by competitors in and outside the country also became an important source of new ideas in this stage. However, none of the technological subcomponents was a significant source of external influence. This is not surprising. Since technology for products (e.g., radio, TV, low price stereo sets, calculators, watches, communication and broadcasting equipment, and simple testing instruments) produced in this industry and their processes was widely and generally known, new technological advances in and outside the country, high technological capability available at local R&D institutions, or systematically indexed information on recent technological development were not significantly related to imitative changes in the industry.

IMPROVEMENT

The electronics industry was initially established in Korea primarily through the international transfer of 'packaged' technology. *Implementation* of the production operation with limited engineering was the immediate technological concern. In the following stage, *assimilation* of foreign technology and product diversification through imitation of foreign products were the important technological tasks necessary to cope with increasing market competition. Then, the relatively successful assimilation of general production technology and increasing emphasis upon export promotion, together with the increased capability of local scientific and engineering personnel, led to gradual *improvement* of foreign technology through local efforts in research, development and engineering (RD&E).

For Korea and many similar countries, whose industrialization programs are based on foreign loans and which have few natural resources to aid repayment and to finance increased imports, export expansion of locally manufactured products was an important alternative available for obtaining the necessary foreign exchange to settle the international balance of payments. For this reason, export promotion has become a national economic development policy for Korea. This forced productive units to enhance their competitiveness in international markets through imitation of more sophisticated foreign products and *improvement* of foreign technology.

This is what is currently taking place in some product lines in the Korean electronics industry.

Although specific 'unpackaged' technologies may be imported from abroad, local RD&E (research, development and engineering) efforts have also become a necessary course of action for local industry not only to improve imported technology, but also to implement their own novel ideas. Thus, research as well as development and engineering gradually became important in this stage. No reliable data are available on the industry wide R&D budget, but productive units under this study showed substantial R&D efforts. 21 out of the 31 productive units had a separate R&D department. These 31 organizations reported an average of 5.28 product changes per year in the past five years. More complete coverage of the innovativeness of these organizations is presented in another report [13].

Productivity improvement and the economy of scale also became important for productive units at this stage in order to be competitive in local and international markets. As indicated previously by the positive correlation between operations technology rigidity and indigenous technological capability, production processes which were generally inefficient in the earlier stages had to gradually become more efficient and integrated. Many productive units attempted to introduce industrial engineering and production management techniques to rationalize production processes. Fig. 5 shows the production scale increase of the electronics industry between 1968 and 1975. The average annual production value per productive unit increased almost four times from US$5.8 million in 1970 to US$22.4 million in 1975. Productivity per worker also increased, as shown in fig. 2, about 2.3 times from US$3,600 in 1970 to US$8,340 in 1975 [11].

Furthermore, successful assimilation of foreign technology through accumulated experience in production and product design and limited efforts in local R&D activity led to the application of imported technologies to different product lines. For instance, a productive unit which imported foreign technologies to produce radio communication equipment designed a custom-ordered electronic quality control device for the government's cigarette plants. The acquisition and assimilation of foreign technologies in communication carrier

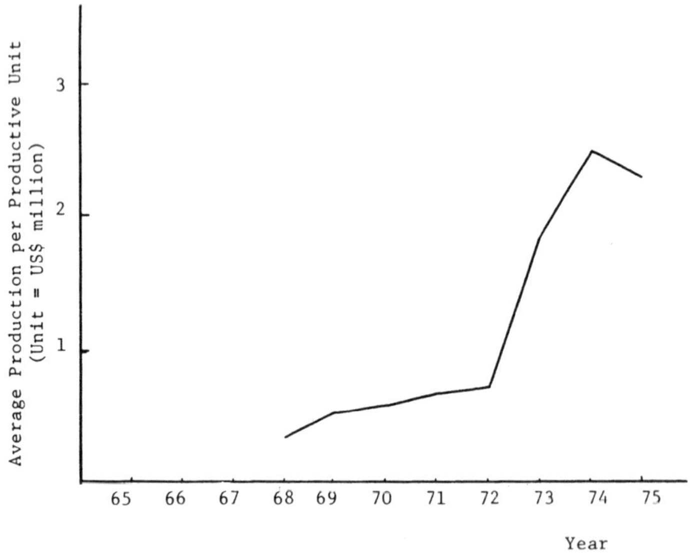

Fig. 5. Average production scale in the electronics industry in Korea, 1968–1975. Source: Table 23 in [11].

equipment led to the production of microwave terminal equipment. Technologies for SSB (single side band) transreceiver and fish finders were applied to the production of broadcasting transmitters of various frequency bands and powers and to radar [15]. Electronic calculator technology and the technical assistance of local applied research institutes are currently leading to the establishment of minicomputer and microprocessor production in Korea. That is, not all the product lines went through the Implementation stage noted earlier. Assimilation of various foreign technologies, increasing capability of local scientific and technical personnel and local R&D efforts provided a basis for general technological development of the industry and for introduction of more sophisticated product lines without transfer of foreign technology. External influences that were increasingly important in the Assimilation stage became, as indicated in table 1, the most important sources of stimuli for technological change in this stage. Changing needs of consumers or users and information from distributors provided crucial market information that led to the creation of new product ideas. The competitor's innovative behavior caused many other firms to follow and adopt similar technologies.

Table 2 summarizes the three stages of development of industrial technology and critical conditions that stand out at different stages.

It is conceivable that successful evolution through the Improvement stage will

Table 2
Summary of the characteristics of three stages

	Stage 1	Stage 2	Stage 3
New firms established through	Transfer of foreign technology	Mobility of local technical and entrepreneurial personnel	
Technical task emphasis	*Implementation* of imported technology	*Assimilation* for product diversification	*Improvement* for enhancing competitiveness
Critical human resources	Foreign experts	Local technical personnel trained at supplier firms	Local scientific and engineering personnel
Production technology	Inefficient		Relatively efficient
Predominant source of technological change	Transfer of 'packaged' foreign technology		Indigenous efforts
Predominant form of international transfer of technology	'Packaged'		'Unpackaged'
Predominant sources of external influence on technological change	Supplier and government		Customer, competitor
Market	Local (low competition)		Local and overseas (high competition)
Emphasis on research, development, engineering	Engineering (E)	Development and Engineering (D&E)	Research, Development, and Engineering, (RD&E)
Supply source of components and parts	Mostly foreign		Mostly local
Important government policies	Import substitution and control of foreign investment		Export promotion
Role of local applied R&D institutions	Consultative	Adaptive development	Research and development

lead Korea to a position where its technological capability will be a challenge to other industrialized countries in the world.

What were initial conditions that made this pattern of development possible in Korea? Several factors seem to have played a significant role in the initial stage. Major factors include, among other things, the government's development policy, availability of local entrepreneurs and skilled technicians and engineers. An aggressive and balanced government development strategy that started with import substitution policy created a protected local market by restricting imports and provided various incentive programs to induce import substitution productions. There were local entrepreneurs who knew how to put foreign capital and technology and local human resources together to exploit a new business opportunities in the protected market. There was also an abundant supply of highly skilled labor force, that quickly learned and assimilated product design and production techniques. That is, aggressive investments in education during the postwar period provided an important base for the subsequent industrialization process. The evidence also indicates that there is a high correlation in the 19th century between educational indices and the speed with which individual Western countries were able to adopt and modify Britain's new industrial technology [22]. It was also the local technical and entrepreneurial capability that facilitated Japan's industrial development.

APPLICABILITY OF THE MODEL

The three-stage model developed above is on the basis of consumer and industrial electronics equipment manufacturing industry in Korea.

However, evidence indicates that the model also describes the development pattern of technology at least in such industries as electrical appliances, agricultural machinery and automobile industries in Korea.

For instance, the automobile industry was first started in 1962 as an assembly operation of semi-assembled Japanese cars. The automobile assembly industry created a market for the autoparts industry,* and import substitution of component parts has reached 38% in 1969, 62% in 1972 and 95% in 1975 [11,21]. With production and design experience accumulated during this period and the development of the local autoparts industry, together with the government's automobile industry promotion programs, the automobile in-

* The autoparts industry existed, in smaller scale, even before 1960 to meet the increasing replacement demand for imported military vehicles and civilian motor vehicles rebuilt from military equipment [28].

dustry introduced in 1974 the first car designed and manufactured in Korea. Furthermore, both the autoparts and automobile industries have recently expanded their markets to overseas. Taking all the evidence together, it can be tentatively concluded that the model may apply to many industries whose production requires a large number of components and parts.

The applicability of the model to process, parts or raw material industries remains to be seen. However, evidence indicates that chemical, pharmaceutical, steel, and synthetic fiber industries were also established through the international transfer of production processes, product specifications, production knowhow, technical personnel and raw material. The growth of these industries led to the emergence of raw material and machinery industries. Subsequent facility expansion and improvement were carried out by local engineers and technicians with local equipment. Furthermore, successful assimilation of imported technology later led to the production of more sophisticated products without foreign assistance.

Is the model also applicable to industries in other countries? Several studies conducted in other countries for different purposes provide useful evidence. Industrialization in Japan evolved through a similar pattern. At the national level, Shishido [23] divided the 100 year history of industrialization in Japan into four steps: (1) the primitive stage which fully relied on imported technology and foreign engineers, (2) the transfer of industrial activity from the government to private sectors, (3) the assimilation of imported technology, and (4) the catching-up process to advanced countries. The industrial technology in the 1860s and onwards was wholly dependent on technology imported from Europe and North America. Technologies imported included not only plant, machinery and raw material, but also foreign engineers and craftsmen.

In the case of textiles, the Japanese government established the first textile mill in 1872 with French technology and engineers. Such technology was diffused quikly within the country. By 1880, Japanese technicians experienced in the government plant were able to play a pivotal role in establishing private spinning mills [23]. The expansion of the textile industry resulted in the emergence of the spinning machine industry in Japan.

The automobile industry in Japan also grew through a similar process. Although attempts were made to develop military trucks after British and French models, the automobile industry in Japan can be said to have started with an assembly operation of American Ford and GM passenger cars in the 1920s. Assimilation of foreign automobile technology through product design and production experience led Japan to introduce its own model 11 years later [27]. Through the process of assimilation and improvement Japan became one of the leading automobile exporters in the world.

Japan's post-war technological recovery was also due to imported technology. Ozawa [20] noted that Japan *licensed* a great number of foreign technologies. Then, its export competitiveness during the 1950s stemmed from a successful *assimilation* of foreign technology. And in the 1960s, further expansion of exports through enhanced competitiveness sprang from the *improvement* of acquired technologies through original research.

To a certain extent, the United States also went through a similar pattern a century ago. The U.S. reversed the sequence of research, development and engineering (RD&E); that is, it started with practical technologies imported from European countries, and then slowly evolved into the present position of being able to afford substantial research [19]. For example, the U.S. imported steam engine technology from the Great Britain in the late 18th century. But through assimilation of British technology, American mechanical skills were well developed by the 1830s and did not need to rely upon or imitate British models and techniques [22].

Baranson [1] also observed that Brazilian firms (excluding foreign subsidiaries) began manufacturing with assembly and packing operations and eventually extended into more intricate operations. Engineering was the initiating portion of the RD&E (research, development and engineering) spectrum leading gradually to more advanced development and research efforts.

DISCUSSION

This article provides a skeletal description of the development pattern of industrial technology as it appears to have taken place in productive units in the electronics industry in Korea. It also provides similar evidence in other industries in Korea and other countries. Many variables both at the industry and national levels may, however, account for variations in the development pattern of industrial technology in different countries. The characteristics of production technology and market structure may be sources of variation at the industry level. Policy on export promotion and foreign investment, the availability of local entrepreneurs, the capability of the local scientific and engineering personnel, and the availability of natural resources may be important sources of variation at the national level. Nevertheless, the general pattern of development of industrial technology that proceeds from *implementation* of imported technology to *assimilation* and eventually to *improvement* thereof in order to strengthen competitiveness is evident in the history of manufacturing industries in different countries. More definite conclusions as to what parameters determine what variations, however, require empirical investigations in different industries and in different LDCs.

As mentioned earlier, this study is a distinct departure from previous studies. It provides a new framework and a base for formulating several tentative propositions, that can be researched in future, on the development process of industrial technology in different LDCs. These propositions include:

–New productive units are established through the transfer of foreign technology in the early period (called Implementation stage) and through the mobility of local technical and entrepreneurial personnel in the intermediate period (called Assimilation stage).

–Implementation of imported foreign technology is most frequently perceived as an important technical task in the Implementation stage, assimilation in the Assimilation stage, and improvement in the Improvement stage.

–International transfer of technology occurs predominantly in the 'packaged' mode in the Implementation stage and in the 'unpackaged' mode in the Assimilation and Improvement stages.

–International transfer of technology is a more frequent source of the generation of new technology than contractual research with local R&D institutions or internal R&D efforts in the Implementation stage, while internal R&D effort is a more frequent source than others in the Improvement stage.

–Government measures to ensure local market protection (e.g., import substitution, control on foreign investment, etc.) are perceived as a most important source of stimulus for the generation of a new technology in the Implementation stage, while government measures to ensure market competition (e.g., export promotion, and anti-monopoly) are perceived as a most important source of stimulus in the Assimilation and Improvement stages.

–Human resources in greatest demand for the generation of a new technology are foreign technical experts dispatched to the recipient productive units in the Implementation stage, local technical personnel trained at the supplier firm in the Assimilation stage, and local scientific and engineering personnel in the Improvement stage.

–Large firms that have previous trade connections are more likely to generate a new technology through the international transfer of technology in the Implementation stage, while small firms with technical entrepreneurs are more likely to generate a new technology through internal efforts in the Improvement stage.

–The degree of export orientation is positively related to the generation of new technology.

–An important role of local R&D institutions is consultation in the Implementation stage, adaptive development in the Assimilation stage, and R&D in the Improvement stage.

–Diffusion of technology is predominantly through the international transfer of technology in the Implementation stage and through the mobility of local technical personnel between firms in the Assimilation and Improvement stages.

–Production process technology is inefficient in the Implementation stage and becomes relatively more efficient in the Assimilation and Improvement stages in response to increasing labor cost pressure.

–Multinational companies that supply capital, raw materials, and technology and Government are the predominant sources of external influence on technological change in the Implementation stage, while customers and competitors are in the Assimilation and Improvement stages.

–Emphasis of local efforts is on engineering (E) in the Implementation stage, development and engineering (D&E) in the Assimilation stage, and research, development and engineering (RD&E) in the Improvement stage.

–Components and parts are mostly foreign in the Implementation stage and mostly local in the Improvement stage.

Most of variables in the above propositions can be measured through examination of records of government agencies, trade association, banks, and other secondary data sources and through interviews with managers. Three stages of technology development can be measured by either one of two surrogate measures: part substitution rate and indigenous technological capability. The study of Nabseth and Ray [18] may be useful to developing a measure for diffusion of technology.

In conclusion, the model presented in this study explains the process of firm's behavior in a LDC in importing, adapting, assimilating, and improving foreign technologies in response to the changing competitive environment and how internal and external factors affect this process. If the model becomes generalizable through subsequent research in different industries in different countries at a similar development stage, it will provide useful implications not only for policy makers and managers of comparable countries, but also for those of other developing countries which attempt to follow the same path and for those of advanced countries in formulating international science and technology policy.

REFERENCES

[1] J. Baranson, *The Drive Toward Technological Self-sufficiency in Developing Countries,* paper presented at the Conference on Latin America-United States

Economic Interactions: Conflict, Accommodation, and Policies for the Future, at the University of Texas at Austin (March 19, 1973).

[2] W. A. Davig, *The Transfer of Production Technology to Brazil: A Study of Two Industrial* Sectors, Ph.D. dissertation at Northwestern University, Evanston, IL (1974).

[3] R. B. Duncan, Characteristics of Organizational Environment and Perceived Environmental Uncertainty, *Administrative Science Quarterly* (September, 1972) 313–327.

[4] Electronics Industry, a Monthly Magazine published by Korean Electronics Manufacturers Association (August, 1975).

[5] FIC (Fine Instrument Center), *Korea Electronics Industry Almanac, 1974* (Fine Instrument Center, Seoul, Korea, 1974).

[6] FIC (Fine Instrument Center), *1975–1976 Directory of Korea Electronics Manufacturers* (Fine Instrument Center, Seoul, Korea, 1975).

[7] M. Goldsmith (ed.), *Technological Innovation and the Economy,* (Wiley Interscience, London, 1970) 3.

[8] D. J. Hickson, D. S. Pugh and D. C. Pheysey, Operations Technology and Organization Structure: An Empirical Reappraisal, *Administrative Science Quarterly* 14 (1969) 378–397.

[9] KDB (Korea Development Bank), *Jeonja Kongup Siltae Josa (Survey of the Electronics Industry),* (Research Division, Korea Development Bank, Seoul, Korea, 1970).

[10] KDB (Korea Development Bank), *Hankuk eui Sanup I, 1973 (Industry of Korea, Volume I, (1973)* (Research Division, Korea Development Bank, Seoul, Korea, 1973).

[11] KDB (Korea Development Bank), *Hankuk eui Sanup I, 1976 (Industry of Korea, Volume I, 1976)* (Research Division, Korea Development Bank, Seoul, Korea, 1976).

[12] L. Kim, *Technological Innovation, Environment, Structure, and Policy Implications: Contingency Approaches,* Unpublished doctoral dissertation, Indiana University (1976).

[13] L. Kim, *Organizational Innovation and Structure: An Empirical Investigation in a Developing Country,* Center for Policy Alternatives, Massachusetts Institute of Technology, Cambridge, MA., CPA/WP–77–1 (1977), *Journal of Business Research* (to appear).

[14] L. Kish, *Survey Sampling* (Wiley, New York, 1965).

[15] KIST (Korea Institute of Science and Technology), *Survey of Technology Imported by Foreign Loan Financed Firms* (Ministry of Science and Technology, Seoul, Korea, 1969).

[16] MCI (Ministry of Commerce and Industry), *Junja Kongup Yuksong Chek kwa Hyunhang (Promotion Policy and Current Situation of the Electronics Industry)* (Ministry of Commerce and Industry, Seoul, Korea, 1974).

[17] M. J. Moravcsik, *Science Development: Toward the Building of Science in Less Developed Countries,* International Development Research Institute, Indiana University (undated).

[18] L. Nabseth and G. F. Ray, *The Diffusion of New Industrial Processes: An International Study* (Cambridge University Press, London, 1974).

[19] NAS (National Academy of Sciences), *U.S. International Firms and R,D,&E in Developing Countries* (National Academy of Sciences, Washington, D.C., 1973).

[20] T. Ozawa, *Japan's Technological Challenge to the West, 1950–1974: Motivation and Accomplishment* (M.I.T. Press, Cambridge, MA, 1974).

[21] Z.H. Park, *The Technical Inducement in the Process of the Industrialization in Korea,* Master's thesis, Sungkyungkwan University, Seoul, Korea (1975).

[22] N. Rosenberg, *Perspectives on Technology* (Cambridge University Press, London, 1976).

[23] T. Shishido, *Japanese Policies for Science and Technology,* Unpublished mimeograph, Nikko Research Center, Tokyo, Japan (1972).

[24] R. D. Steele, *Multinational Corporations, Technology Transfer, and the Developing Countries: An Introductory Bibliography,* Monticello, Il, Council of Planning Librarians (1975).

[25] W. S. Tae, *Development of Korean Economy: Past, Present, and Future* (Samwha Publishing Company, Seoul, Korea, 1972).

[26] World Bank, *World Tables, 1976* (Johns Hopkins University Press, Baltimore, MD, 1976).

[27] T. Yakushiji, *Dynamics of Policy Interventions: Government and Automobile Industry in Japan, 1900–1960,* Ph.D. dissertation M.I.T. Cambridge, MA (1977).

[28] J. M. Yoon, *Jadongcha Kongup 70 Nyunsa (70 Year History of Korean Automobile Industry)* (Korea Automobile Industry Association, Seoul, Korea, 1975).

MANAGEMENT SCIENCE
Vol. 31, No. 3, March 1985
Printed in U.S.A.

INNOVATION IN A NEWLY INDUSTRIALIZING COUNTRY: A MULTIPLE DISCRIMINANT ANALYSIS*

LINSU KIM AND YOUNGBAE KIM
Korea Advanced Institute of Science and Technology, P.O.B. 150, Cheongryang, Seoul, Korea

This article examines 42 innovative firms in a newly industrializing country, Korea. It identifies different patterns of innovation associated with four types of firms, which is categorized by two variables: the source of initiation and the local availability of related foreign products. Bivariate analyses suggest that innovation patterns are distinctly different for Type I firms (user initiated and related foreign products locally available) and Type IV firms (innovator initiated and no related foreign products locally available). Type I firms use and benefit most from close interactions with customers and visits to the domestic users of related foreign products, while Type IV firms benefit most from the overseas observation of foreign suppliers and the assistance of local R&D institutes. The other two groups are more closely related but still different from the above two groups. The results of a multiple discriminant analysis are supportive of the patterns which emerged in the bivariate analyses. It was also found that majority of important and crucial information to solve technical problems were transferred free of charge from abroad through informal mechanisms rather than formal collaboration with foreign firms. The findings suggest that different situations call for different approaches to make innovations successful.
(RESEARCH AND DEVELOPMENT; TECHNOLOGY TRANSFER; ECONOMIC DEVELOPMENT)

Many developing countries have made significant progress in industrial and technological development, evolving from agriculture-dominant traditional societies to industry-dominant modern societies. In the process, the management of technological change has become increasingly important. This raises two questions: (1) Can theories pertinent to the innovation process in developed countries (DCs) be generalized to developing countries? (2) How and why is the pattern of the innovation process in developing countries different from that of DCs?

There are two principal sources of obtaining technology for firms in both newly industrializing countries (NICs) and less developed countries (LDCs);[1] first, from abroad through multinational companies' direct foreign investment, the purchase of turnkey plants, patent and knowhow licenses, and technical services; second, locally through research and development, copying or reverse engineering of foreign products and processes, and self-directed efforts to improve and adapt previously acquired technology. The second source requires a significant degree of local capability. In the case of NICs, the second approach becomes increasingly important in order to strengthen its international competitiveness in the face of the increasing threat from LDCs which strive aggressively to develop labor intensive light industries with lower wages and to penetrate into relatively more technology-based industries that have been

*Accepted by Burton V. Dean; received March 9, 1983. This paper has been with the authors $1\frac{1}{2}$ months for 1 revision.

[1]Grossly, developed countries refer to member countries of OECD, except Greece, Portugal and Turkey. This definition is consistent with World Bank definition of industrial market economies (World Bank 1982). Newly industrializing countries refer to those countries which are small in number, export-oriented, and have achieved very rapid rate of economic growth through the expansion of industrial sectors over the past decade. Among them are Brazil, Korea, Singapore, Taiwan, and Hong Kong (Krueger 1982). Less developed countries usually refer to other developing countries that strive to follow the same path. Oil exporting and nonmarket economies are not included in this categorization.

0025-1909/85/3103/0312$01.25

dominated by firms from DCs. In addition, firms in DCs show increasing reluctance to license technologies to NICs that have a potential to become competitors in the international market. For these reasons, indigenous efforts for innovation become important in NICs to sustain competitiveness in the changing international market. An earlier study showed that technological innovation through local effort was perceived by firms in a NIC as one of the three most important strategic means to survive and grow in increasingly competitive environments (Kim 1980a). This article examines the pattern of process associated with locally developed innovations in a newly industrializing country, Korea.

Studies of the process of technological innovation are concentrated mainly in the US and UK (Utterback 1974). But the findings of these studies cannot necessarily be generalized to both NICs and LDCs. Little research of this type has been done in newly industrializing countries. A few exceptions might include studies conducted in Latin America (Katz 1980), Taiwan (Gerstenfeld and Wortzel 1977), and Korea (Kim 1980b). These studies either focus primarily on the process of assimilating imported technology or test research frameworks developed in DCs.

As a departure from previous studies, this article attempts to identify, on the basis of a new research scheme, different patterns in firms' innovation behavior in a newly industrializing country. It also assesses the role of informal technology transfer across nations in indigenous technological efforts in NICs.

With the recent erosion of comparative advantage of cheap skilled labor in NICs, a demand has steadily increased for new capital goods that can lead to quality improvement and cost reduction. Due to the high cost of foreign imports and tariff and nontariff barriers, many capital goods users approach local firms for machines which are less expensive but more appropriate to the prevailing local conditions than foreign imports. This results in important innovations in the machinery industry. These innovations are, indeed, market pulled (triggered by either user's orders or serious inquiries). But to avoid confusion with the much criticized concept of "market pull and technology push" (Mowrey and Rosenberg 1979), the term "user initiated" is coined here to describe the above phenomena.[2] There are also cases where innovative firms take initiatives and either respond to direct market pressure (passive) or deliberately search for new markets and techniques to make their ideas commercially and technically successful (active). In this case, innovations are "innovator initiated."[3]

The source of initiation (user initiated versus innovator initated) appears to be an important variable to understand difference in the innovation behavior of firms. Firms (or project teams) that respond to specific orders or inquiries from users are likely to have better ideas about users' needs, but less information about technical feasibility, while firms that deliberately search for new markets and techniques for innovation are likely to have less information about potential users and the marketability of the ideas. But they are apt to have better ideas as to how a technically feasible product can be developed, because technical feasibility must have been considered at the start of the project. Accordingly, their information searching and problem solving behaviors plus communication patterns with markets and technical sources associated with user initiated innovations are expected to differ significantly from those with innovator initated innovations.

[2] As noted by von Hippel (1976, 1977, 1978) and Pavitt and Walker (1976), users' role in innovation also appears to be dominant in the machinery industry in Korea, in which this study was undertaken. But unlike industries reported in von Hippel's studies, the users in NICs are not important initial developers of new innovations; rather they are important initial sources of stimulus either through order placing or through serious inquiries.

[3] Carter and Williams' (1959) passive innovations also include user initiated cases. Therefore, our classification is not consistent with any of the existing ones.

Most innovations in newly industrializing countries, though significant in a local sense, are not original in the global sense. Firms in NICs learn from related foreign products, if they are locally available. For this reason, *the local availability of related foreign products* (hereinafter referred to as RFPs) may be an important variable, when identifying differences in firms' behavior with respect to innovation in NICs. Firms that have RFPs locally available can learn to a great extent from them, while firms without them have to look for different means. Correlating these two variables—the source of initiation and the local availability of related foreign products, a research scheme of four cells is identified: (1) user initiated—RFPs locally available, (2) innovator initiated—RFPs locally available, (3) user initiated—no RFPs locally available, and (4) innovator initiated—no RFPs locally available. The following hypothesis reflects this scheme.

Hypothesis. Four different types of firms in NICs, categorized by the source of initiation and the local availability of related foreign products, exhibit different patterns of innovation.

Studies in DCs suggest that the process of developing new products or processes may be conceived as consisting of several sequential stages (Myers and Marquis 1969; Utterback 1971; Zaltman et al. 1973; etc.). There are noticeable variations among these stage models, but most of them imply three: idea generation, problem solving, and implementation. These stages, developed on the basis of studies undertaken in DCs, do not necessarily reflect events in NICs, nor do they always occur in a linear sequence. However, they are still useful for examining the innovation process in NICs. An arbitrary but useful division of the first two stages is the point in time at which a decision is made to expend resources for the development of a solution. The idea generation for an innovation involves, among other things, a collection of both internal and external information to estimate the probability of technical success, the cost of development, the probability of commercial success, and the expected return to the firm. The problem solving stage in this study includes both problem solving activity and solution.

In the idea generation stage, both internal and external agents such as key persons within the firm and users in the market and environmental factors such as import substitution policy can play important roles. In the problem solving stage, key persons within the firm, interactions with external agents such as users and local research institutes, the observation of related foreign products within the country and abroad, and printed technical information such as catalogs could play important roles. The innovation pattern associated with these variables is likely to differ significantly by the four cell scheme presented above. Finally, the patentability of innovations and the degree of imitation might also differ by the four cells. These variables are included in the study to test the aforementioned hypothesis.

Methods

Sample

The data for this study were collected from 42 innovative firms in the machinery industry in Korea, one of the most rapidly industrializing countries in the world. Growth in real GNP for the past two decades was more than 4.9 fold, reflecting an average annual growth rate of 8.3%. Its manufacturing sector grew faster during the same period at the annual rate of 16.5%. The machinery industry is one of its fastest growing industries; its share within the manufacturing sector increased from 6.3% in 1960 to over 20% in 1980.

The 42 firms were selected from the population of 70 in the machinery industry.

These firms have developed significantly innovative products without foreign collaboration and have been consequently designated by the Korean government as producers of "newly developed innovative machines" (hereinafter referred to as NDIMs). In 1979 the Korean government introduced a policy designed to facilitate the development of industrial technology in the industry. Specifically, in 1979 and 1980 the government designated 86 capital goods developed by the 70 firms as NDIMs and offered special incentives to the producers and to those who purchased them. The criteria used in designating these firms include that they should have: (1) been developed with local patents or with a new technology; (2) a domestic parts ratio of over 60%; (3) been developed for the first time in Korea without any technical collaboration from abroad; (4) quality comparable with or better than foreign products as determined by a government R&D institute; (5) originality and not an exact copy of a foreign product; and (6) used domestic parts only for critical functions. The government accepts applications twice a year from firms for NDIM designation. Among the first applicants, only about 20% passed the stringent criteria. The criteria plus the thorough evaluation of applications by the government indicate that those designated contain a significant degree of local innovativeness.

Approach

A pilot study of five firms was first undertaken to test an instrument developed on the basis of the existing literature. The revised instrument was sent to all 70 firms, which had been designated as NDIM producers by the government. Forty-two firms responded. Personal interviews followed the questionnaires in order to verify replies and also to collect more in-depth information. Table 1 presents the distribution of samples arranged by industrial classification and size. It shows that nonelectrical machinery accounts for over 70% and that small firms with less than 100 workers account for over 64%. This distribution is similar to that of the total population of 70 firms.[4] The innovative products included in this study range from an imitative automatic loom to a very indigenous high pressure brick machine.

Using the research scheme presented earlier, the sample was divided into four cells: Type I (user initiated and RFPs locally available), Type II (innovator initiated and

TABLE 1
Sample Distribution by Industrial Classification and Size

Sample Firms			Size (Number of Workers)			
Industry (ISIC 38)	*n*	%	1 – 99	100–199	200–499	500 or More
Metal fabrication	1	2.4		1		
Nonelectrical machinery	30	71.4	20	2	4	4
Electrical machinery	5	11.9	3	1		1
Transport machinery	4	9.5	4			
Precision machinery	2	4.8		1	1	
Total	42	100.0	27	5	5	5

[4]The distribution of the population shows that nonelectrical machinery accounts for 68.8%, electrical machinery for 11.4%, transport machinery for 8.6%, etc. and firms with less than 100 workers account for 62.9%, firms with workers 100–499 account for 27.1% compared to 23.8% in the sample, firms with workers over 500 account for 10.0% compared with 11.9% in the sample.

TABLE 2
Research Scheme and Sample Distribution

		Source of Initiation		
		User	Innovator	Total
Related foreign products locally available from which local firms can learn and creatively imitate?	Yes	Type I $n = 21$	Type II $n = 8$	29 70.8%
	No	Type III $n = 6$	Type IV $n = 6$	12 29.2%
Total		21 65.9%	14 34.1%	41[a] 100.0%

[a] One firm has a missing value.

RFPs locally available), Type III (user initiated and no RFPs locally available), and Type IV (innovator initiated and no RFPs locally available). The two-by-two classification and sample distribution are shown in Table 2.

Results

Table 3 presents the bivariate relationships between the four types of firm and independent variables related to the innovation process. The first block of the table reports the results of cross tabulation for nominal variables, showing, for example, that in 59.1% of Type I firms (the first figure in the table) entrepreneurs were key persons in idea generation. The frequency distributions in the four cell cross tabulation do not violate the requirements suggested by Cochran (1954). The second block reports the results of Kruskal–Wallis ANOVA for interval variables. The results are almost identical to those of parametric ANOVA.

Table 3 indicates that in the idea generation stage there is no difference in the role of entrepreneurs as key persons in initiating an innovation: in any group 60 to 70 percent of firms reported that it was the entrepreneurs that contributed significantly to the initiation. In contrast, Type I and II groups (firms with RFPs locally available) showed higher scores than the other two groups (firms with no RFPs locally available) in the role of customers in idea generation (24.64 and 15.00 compared with 12.33 and 11.90). The result of K–W ANOVA for import substitution policy is not statistically significant, but mean ranks are also higher in Type I and II groups (firms with RFPs locally available) than the other two groups (23.52 and 21.00 compared with 15.17 and 14.67). In other words, when RFPs are locally available, their users play a significant role as external agents in helping generate ideas for innovations in the face of government's import substitution policy.

Table 3 also shows that in the problem solving stage the proportion of entrepreneurs who played the key technical role is significantly higher in Types III and IV groups (firms with no RFPs locally available). That is, entrepreneurs played a central role from the generation of the initial idea to the end of the problem solving when there were no related foreign products to learn from and as a result the development task demanded greater creativeness. It can be inferred that the complete commitment of entrepreneurs must have contributed significantly to the successful innovation for these firms, as several studies in developed countries suggest. For example, Rothwell (1977) and Langrish et al. (1972) found that the lack of commitment by top management is one of the major reasons for failure of innovation.

Table 3 also shows that Type II and IV groups (firms with innovator initiated innovations) reported higher scores than the other two groups (firms with user initiated

TABLE 3

Difference in Innovation Behavior for Four Types

Factors Affecting Innovation	Related Foreign Products Locally Available: User Initiated Type I $n = 21$	Related Foreign Products Locally Available: Innovator Initiated Type II $n = 8$	No Related Foreign Products Locally Available: User Initiated Type III $n = 6$	No Related Foreign Products Locally Available: Innovator Initiated Type IV $n = 6$
(Results of Cross Tabulations—Percentage)[a]				
Entrepreneurs as:				
Key person in idea generation	59.1	71.4	66.7	66.7
Key person in problem solving	36.4	28.6	50.0	83.3
Firms used overseas observation	19.1	25.0	16.7	33.3
Firms obtained patents for NDIMs	5.0	12.5	16.7	66.7
(Results of Kruskal–Wallis ANOVA—Mean Ranks)[b]				
Educational level of key technical person	17.56	15.00	18.67	26.60
Position of key technical person within the firm	15.97	14.75	19.92	23.10
Customer's role in:				
idea generation***	24.64	15.00	12.33	11.90
problem solving***	26.02	18.05	14.83	7.30
Opportunity opened by import substitution policy	23.52	21.00	15.17	14.67
Contribution of domestic observation**	24.48	22.57	13.42	11.25
Firms obtained assistance from local R&D lab***	18.95	20.50	18.00	31.83
Usefulness of catalogs*	22.63	20.29	18.42	9.75
Degree of imitation***	27.42	16.25	13.92	9.67

[a] The figures reported in the first block is the proportion of firms that responded "yes". For example, 59.1% (first figure in the table) of the firms (Type I), whose innovation was triggered by users and who had related foreign products locally available, mentioned that it was the entrepreneur that played a key role in the initiation of innovation.

[b] *significance level < 0.1.

**significance level < 0.05.

***significance level < 0.01.

innovations) in both the use of overseas observation of foreign firms producing related products (25.0 and 33.3 compared with 19.1 and 16.7) and the use of assistance from local R&D institutes (20.50 and 31.83 compared with 18.95 and 18.00). In contrast, Type I and II groups (firms with related foreign products locally available) reported higher scores than the other two groups (firms with no RFPs locally available) in the role of customers in technical problem solving (26.02 and 18.05 compared with 14.83

and 7.30), in contribution of domestic observation of local firms using related foreign products (24.48 and 22.57 versus 13.42 and 11.25), and in the use of catalogs (22.63 and 20.29 compared with 18.42 and 9.75). As a result, innovations in Type I group (user-initiated and related foreign products locally available) were most imitative and those in Type IV group (innovator-initiated and no related foreign products locally available) least imitative. Consequently, 66.7% of innovations in Type IV group were patented compared with only 5% in Type I group.

Contrasts were noted with the most extreme differences occurring between Types I and IV. When the innovation was triggered by either users' orders or serious inquiries and there were related foreign products locally available from which firms could learn and imitate creatively, firms used and benefited most from close interactions with users in both the idea generation and problem solving stages and visits to the domestic users of such foreign products plus catalogs readily available at the local office of foreign suppliers. Consequently, such innovations were most imitative and least patented. In contrast, when the innovation was initiated by innovators searching for markets and techniques and there were no related foreign products locally available from which firms could learn, they used and benefited most from overseas observation of foreign suppliers and assistance of local R & D institutes. They also required higher quality key technical persons within the firm. As a result, such innovations were least imitative and most patented.

Next, a multiple discriminant analysis was undertaken to see if interval variables used for bivariate analyses significantly differentiate the four groups in a multivariate setting. One may argue that the sample size of 42 with four groups may not be enough to assume multivariate normality in distribution, but this technique is not particularly

TABLE 4

Results of Discriminant Analysis
(Stepwise)

Canonical Discriminant Functions				
Function	Eigenvalue	% of Variance	Cumulative	Canonical Correlation
1	2.79	81.43	81.43	0.8579
2	0.45	13.08	94.51	0.5561
3	0.19	5.49	100.00	0.3977
After Function	Wilks Lambda	Chi-squared	D.F.	Significance
0	0.1535	64.64	21	0.0000
1	0.5815	18.70	12	0.0960
2	0.8418	5.93	5	0.3122

Standardized Discriminant Function Coefficients ($n = 41$)

Variables	Function 1	Function 2
Import substitution	0.58357	0.35391
Domestic observation	0.17986	− 0.22145
Education of key technical person	− 0.29291	− 0.40600
Role of users in idea generation	0.37942	− 0.45163
Position of key technical person within the firm	− 0.67187	− 0.34195
Contribution of local R & D institutes	− 0.82488	− 0.57567
Degree of imitation	0.33960	− 0.56150

sensitive to minor violations of the assumption (Klecka 1980). Table 4 summarizes the results of stepwise discriminant analysis. Seven of the original nine variables were selected when Wilk's lambda reached the lowest level. The first function produced a very high degree of separation among the four groups as indicated by the final Wilk's lambda (0.1535) and canonical correlation (0.8578). The second function is not as powerful but is useful. Since the third function did not add any more discriminating power in the analysis, the first two functions were used to discriminate the four groups. Standardized canonical discriminant function coefficients for the two functions are also reported in Table 4.

The two-function plot of group centroids and individual cases, as presented in Figure 1, supports the general picture of bivariate analyses in Table 3. It shows that the centroids of Types I and IV are well separated not only with each other but also from those of Types II and III, which are more closely located, indicating that Types I and IV are quite distinct, while the other two are less so. However, individual cases are well clustered around the group centroids, except for Type II. The location of the four centroids also indicates that the first function discriminates all four groups, but the second function discriminates Type IV from the rest.

The predictive results of the discriminant analysis, as presented in Table 5, show that 80.49% of firms under study were correctly classified. As expected from the above plot, over 90% of Type I firms and 100% of Type IV firms were correctly classified. The proportion was about 50 to 60 percent for Types II and III. One may argue that to test the usefulness of the results of the discriminant functions derived from the analysis they should be applied to samples not including 42 firms from which the functions originated. However, the focus of the analysis here is not to predict firms' group membership on the basis of values on the discriminant variables. Rather, its aim is at determining variables that are systematically associated with the four groups of firms. In conclusion, the results of the multivariate discriminant analysis largely support those of bivariate analyses and in turn support the hypotheses that firms in the four groups (at least three groups) exhibit different patterns of innovation behavior.

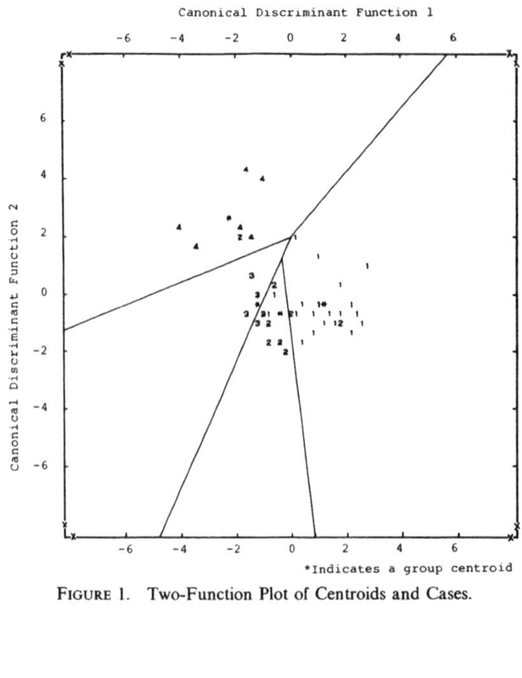

FIGURE 1. Two-Function Plot of Centroids and Cases.

TABLE 5

Prediction Results of Group Classification

Actual Group	No. of Cases	Predicted Group Membership I	II	III	IV
Type I	21	19 90.5%	2 9.5%	0	0
Type II	8	2 25.0%	5 62.5%	0	1 12.5%
Type III	6	0	3 50.0%	3 50.0%	0
Type IV	6	0	0	0	6 100.0%

Percentage of grouped cases classified: 80.49%.

As the data presented in our tables suggest, the informal transfer of foreign technology by foreign machinery producers plays an important role in local innovations. For example, first they sell their products to local users, from which local innovators can learn and creatively imitate. Second, foreign suppliers release catalogs and other technical information hoping to be able to sell their products. Our sample indicated that 87% of the firms interviewed collected catalogs of related foreign products from local agents or directly from foreign producers. From this group 94% of those who obtained such catalogs replied that they had benefited from them. Third, when a related foreign product was not locally available, firms sent key technical personnel to foreign producers to examine products and obtain catalogs and product specifications. Ten out of 42 firms mentioned that they had benefited to a great extent from such visits to foreign suppliers. In other words, the majority of important or crucial information to solve technical problems were provided free of charge through informal mechanisms rather than through formal collaborations.

Summary and Discussion

Most studies on the innovation process are concentrated mainly in the US and UK, but the findings of these studies cannot necessarily be generalized to developing countries. Little research of this type has been done in newly industrializing countries which have increasing concern about innovation. This study attempted to identify different patterns of innovation behaviors in local firms in Korea that have produced significantly innovative products without formal collaborations with foreign firms. The findings suggest that factors associated with innovations are significantly different among four groups of firms categorized by two variables: the source of initiation and the local availability of related foreign products. These findings offer several implications.

First, the management of local firms should take a contingent approach to make indigenous innovations successful. For example, if a project is triggered by a user's order or inquiry and a related foreign product is locally available, then the firm should exert its efforts to develop effective contacts with users who placed the order or inquiry and the firms that use a related foreign product. They can provide the most critical information necessary for the success of the project. Should an innovation project be initiated by the innovator and with no related foreign products locally available, the top management should give full commitment to the project, develop an effective relationship with local science and technology infrastructure, and send key technical persons to visit foreign producers.

Second, this study, though original in its conceptualization, is much indebted to studies undertaken in developed countries. This fact implies that research on technological innovation in both newly industrializing and less developed countries can benefit from knowledge generated in developed countries. However, research can be significantly more creative and original when the local situation is taken into consideration in research design. Research should be both deductive, based on existing literature from developed countries, and inductive, based on careful observation of local situations.

Third, the importance of technology transfer in the innovation process is well recognized (Gee 1974). In the newly industrializing country the transfer of foreign technology that has been tested and proven in the DC markets can reduce risks and costs, and shorten the time span of generating new products. But our knowledge of international technology transfer is much limited (Lee and Kim 1982). Most studies on the issue are from perspectives of developed countries, concerned mainly with technology transfer through the mechanisms of equity participation and contractual agreements such as licensing and know-how contracts (e.g., Teece 1976; Michalet 1976; Berman 1976; Baranson 1978). All these mechanisms require sophisticated bargaining techniques and financial resources, which only large firms in NICs and LDCs appear to possess. There are few studies that examine other mechanisms that are less expensive and that can easily be used even by small and medium-sized firms. In contrast, the findings of this study suggest that informal mechanisms play a significant role in transferring important technology from DCs to NICs. Future studies should shed more light on the informal mechanism.

Finally, it is necessary to replicate this study in different environments. The pattern observed in Korea which has adopted conservative policies toward foreign investors does not necessarily prevail in other newly industrializing countries such as Brazil and Mexico, where liberal policies toward foreign investments by multinational companies have been adopted. Even within one country different patterns may be observed in different industries. For example, consumer electronics firms can easily purchase related foreign products in the international market; therefore, their idea generation and problem solving patterns are expected to differ from the findings reported in this study.[5]

[5]The research underlying this article was supported by the World Employment Programme of the International Labour Office, Geneva. The opinions, findings, conclusions, and recommendations expressed herein are those of the authors and do not necessarily reflect the views of the International Labour Office.

References

BARANSON, JACK, *Technology and the Multinationals*, Lexington Books, Lexington, Mass., 1978.

BERMAN, JACK N., *Transfer of Manufacturing Technology within Multinational Enterprises*, Ballinger, Cambridge, Mass., 1976.

CARTER, C. F. AND B. R. WILLIAMS, *Investment in Innovation*, Oxford University Press, London, 1959.

COCHRAN, W. G., "Some Methods for Strengthening the Common X^2 Tests," *Biometrics*, 10 (1954), 417–451.

GEE, SHERMAN, "The Role of Technology Transfer in Innovation," *Research Management*, 17, 6 (November 1974), 31–36.

GERSTENFELD, ARTHUR AND L. H. WORTZEL, "Strategies for Innovation in Developing Countries," *Sloan Management Rev.*, 19, 1 (Fall 1977), 57–68.

KATZ, JORGE, "Domestic Technology Generation in LDCs: A Review of Research Findings," Research Programme on Scientific and Technological Development in Latin America, Buenos Aires, Working Paper No. 35, November 1980.

KLECKA, WILLIAM R., *Discriminant Analysis*, Sage University Press, Beverley Hills, Cal., 1980.

KIM, LINSU, "Organizational Innovation and Structure," *J. Business Res.*, 8, 2 (1980a), 225–245.

———, "Stages of Development of Industrial Technology in a Developing Country: A Model," *Research Policy*, 9, 3 (1980b), 154–177.

KRUEGER, ANNE O., "Newly Industrializing Economies," *Economic Impact*, 40, 4 (1982), 26–32.

LANGRISH, J. ET AL., *Wealth from Knowledge*, McMillan Press, New York, 1972.

LEE, JINJOO AND LINSU KIM, "Technology Transfer: A Conceptual Framework for International Comparative Study in Developing Countries," A Report Submitted to UNESCO, December 1982.

MICHALET, CHARLES-ALBERT, "International Transfer of Technology and the Multinational Enterprises," *Internat. Organization*, 7, 2 (April 1976).

MOWREY, DAVID AND NATHAN ROSENBERG, "The Influence of Market Demand upon Innovation: A Critical Review of Some Recent Empirical Studies," *Research Policy*, 8 (1979), 102–153.

MYERS, SUMNER AND DONALD G. MARQUIS, *Successful Commercial Innovations*, National Science Foundation, Washington, DC, NSF 69–71, 1969.

PAVITT, KEITH AND W. WALKER, "Government Policies Towards Industrial Innovation," *Research Policy*, 5 (1976), 11–97.

ROTHWELL, R., "The Characteristics of Successful Innovators and Technically Progressive Firms (with Some Comments on Innovation Research)," *R&D Managment*, 7, 3 (1977), 191–206.

TEECE, DAVID J., *The Multinational Corporation and the Resource Costs of International Technology Transfer*, Ballinger Publishing, Cambridge, Mass., 1976.

UTTERBACK, JAMES M., "The Process of Innovation: A Review of Some Recent Findings," in George Wilson (Ed.), *Technological Development and Economic Growth*, Indiana University Graduate School of Business, Bloomington, 1971.

———, "Innovation in Industry and Diffusion of Technology," *Science*, 183 (1974), 620–626.

VON HIPPEL, ERIC A., "The Dominant Role of Users in the Scientific Instrument Innovation Process," *Research Policy*, 5, 3 (1976), 212–239.

———, "Transferring Process Equipment Innovations from User-Innovators to Equipment Manufacturing Firms," *R&D Management*, 8, 1 (1977), 13–22.

———, "A Consumer-Active Paradigm for Industrial Product Idea Generation," *Research Policy*, 7 (1978), 240–266.

WORLD BANK, *World Development Report 1982*, Oxford University Press, New York, 1982.

ZALTMAN, GERALD, ROBERT B. DUNCAN AND JONNY HOLBEK, *Innovation and Organizations*, Wiley and Sons, New York, 1973.

[3]

Reflections on The Republic of Korea's Acquisition of Technological Capability

Larry E. Westphal,
Linsu Kim, and
Carl J. Dahlman

INTRODUCTION

South Korean industrialization is unusual not only in its remarkable pace but in other significant respects as well. It thus is a fruitful area for studying industrial development, as is demonstrated by what has been written on the trade aspect of the Republic of Korea's industrial strategy. But there has been little study of – and even less written about – the technological development accompanying its industrialization. Our purpose in this chapter is to document the technological aspect of South Korean industrial strategy.

In our previous attempts to convey what we have learned about South Korea's technological development to colleagues, we have been frustrated by the fact that economists typically do not think

---

The research underlying this chapter forms part of an ongoing project, "The Acquisition of Technological Capability," sponsored by the World Bank (Ref. No. 672-48). Westphal had the primary responsibility for writing the chapter. Kim worked mainly on the last three sections; financial support for research assistance from the Social Science Research Council enabled him to undertake the supplementary case history research that is reported in the section "Exports in Relation to Technological Development." Dahlman worked mainly on the first two sections. Bruce Ross-Larson edited the manuscript for publication.

This chapter had its origin in a related paper presented at the Conference on International Technology Transfer: Concepts, Measures, and Comparisons; sponsored by the Social Science Research Council, Subcommittee on Science

of industrialization as technological development. That may be the result of the absence of a satisfactory analytical framework for thinking in these terms. We thus find it necessary to establish a conceptual foundation for discussing the South Korean experience. The first two sections provide this foundation, which – though largely derived from our reflections on South Korea – we believe is more generally valid. South Korea's technological development is described in the next section. Because of the importance of the export-led character of its industrialization, we explore then the interactions between technological development and export activity. Findings and speculations are summarized in the final section.

TECHNOLOGICAL DEVELOPMENT IN RELATION TO INDUSTRIALIZATION

Why is it that economists typically do not associate technological development with the industrialization of developing countries? Perhaps because invention, the central aspect of global technological development, plays only a minor part in the process. Most technology introduced into developing countries is transferred in one way or another from industrially more advanced countries. But because industrialization adds to the variety of products produced and processes used in a country, it surely does involve technological development in the sense of gaining mastery over products and processes that are new to the local economy. The minor role of invention simply means that much technological development consists of assimilating foreign technology.

Another possible explanation for the neglect of technological development in industrialization is that assimilation often seems to

---

and Technology Indicators. We are grateful to the participants at the conference for useful comments and discussion and want to acknowledge a special debt to Richard Nelson. We also wish to acknowledge Alice Amsden's contributions as co-author (with Westphal and Kim) of the original paper. Scheduling conflicts precluded her collaborating on this chapter.

The authors are solely responsible for the views and interpretations expressed in this paper. In particular, these views and interpretations should not be attributed to the World Bank, to its affiliated institutions, or to any individual acting on behalf of these organizations.

be characterized as being automatic and without cost. If this were correct, assimilation would not merit much attention. But it is not accomplished by passively receiving technology from overseas. It requires investments in understanding the principles and use of technology, investments reflected in increased human and institutional capital. These have a big effect on productivity. And they have a big effect on possibilities for adapting technology to suit local circumstances better.

A third possible explanation for the scant attention paid to technological development is that assimilation is seen as being not only absolute but also unitary and discrete. But rather than being absolute, assimilation has an aspect of proficiency. And in addition, the choices associated with investments in assimilation have an aspect of complexity that follows from the character of technology as a compound system of interrelated elements. If this complexity is forgotten, as it often appears to be, the nature of assimilation is trivialized.

The complexity of choices associated with investments in assimilation – complexity inherent in the separability of technology into its many elements – has extremely important implications. Most important is that a country does not have to possess all the capabilities needed to provide each of the various elements of technology. All the elements of technology can in principle be obtained – individually or in a variety of combinations – from foreign sources. It is in this respect that South Korea provides a particularly interesting and instructive case. Its experience during the 1960s and 1970s shows how a country can successfully assimilate technology in piecemeal fashion.

International trade in the elements of technology means, too, that there is no necessary relationship between industrial development and technological development – the former seen in the size and composition of industrial output and the latter in the local capability to provide various elements of technology. Certainly there is no necessary relationship over short periods. That is why many paths of local technological development can be taken to reach the same level of industrialization. Those paths can differ in the mode of technology transfers as well as in the extent and nature of investments to assimilate technology – to acquire the capabilities to provide the various elements. There naturally are differences in costs and benefits for different modes of transfer. But these may be of secondary importance, since they can be overcome by tailoring investments in assimilation to offset them.

How, then, is technological development to be assessed in relation to industrial development? Any survey of a country's past technological development can at best yield imprecise results because of the limited usefulness of readily available information. Many indirect modes of technology transfer, such as ideas conveyed through the open technical literature, are not easily traced. Many forms of investment in assimilating technology, such as those associated with the accumulation of experience, also go unrecorded. Except in rare circumstances, costs cannot be assessed precisely, even through detailed analyses of case histories of firms and sectors. Benefits are at least as difficult to assess because of the difficulties created by the likelihood of externalities. Contemporary attempts to uncover the costs and benefits associated with different paths of technological development are further hampered by the paucity of adequate case histories.

In sum, current surveys of technological development in industrializing countries can only address descriptive questions: What elements of technology have been assimilated and in what sequence? Where did they come from? How were they first obtained and then absorbed? Which of them were adapted, in what respects, and how? These are the questions we address in this chapter.

THE NATURE OF TECHNOLOGICAL CAPABILITY[1]

Analysis of technological development requires a conceptual framework that distinguishes the possible choices associated with investments in assimilation. In building such a framework, we have found it most useful to focus not on technology but on technological capabilities distinguished by the functions served. In this section we will provide a functional classification of technological capabilities and discuss how these relate to trade in the elements of technology.

The term *technology* refers to a collection of physical processes that transform inputs into outputs, to the specifications of the inputs and the outputs, and to the social arrangements that structure the activities involved in carrying out these transformations. Technology thus is the practical application of technological knowledge, which is what underlies and is given usable expression in technology. Such knowledge has technical and transactional elements, the former relating to product characteristics and physical processes, and the latter to social arrangements. These social arrangements include various kinds of market and contractual relationships among entities as well as organizational modes and procedural methods to regulate an entity's internal operations.

Technological capability is the ability to make effective use of technological knowledge. It is the primary attribute of human and institutional capital.* It inheres not in the knowledge that is possessed but in the use of that knowledge and in the proficiency of its use in production, investment, and innovation.† Because of the complexity of technology, there are many distinct technological capabilities, classifiable in numerous ways, each corresponding to a different way of distinguishing the aspects of technological knowledge and its application. We sketch here a classification by broad areas of application.

Functional Classification of Capabilities

Technological capabilities are separable into three broad areas: production, investment, and innovation.‡ The first capability is for

---

**Institutional capital* refers to the know-how used to combine human skills and physical capital into systems for delivering want-satisfying products. In terms of Hall and Johnson's conceptual framework, institutional capital includes system-specific and firm-specific technology plus some elements of general technology. G. R. Hall and R. E. Johnson, "Transfers of United States Aerospace Technology to Japan," in *The Technology Factor in International Trade*, ed. Raymond Vernon (New York: National Bureau of Economic Research, distributed by Columbia University Press, 1970), pp. 305-58.

†The identification of technological capability as something distinct and worthy of attention conforms to Salter's separation of technological knowledge and technology. He uses the terms *technical knowledge* and *techniques of production*. W. E. G. Salter, *Productivity and Technical Change* (Cambridge: Cambridge University Press, 1960), pp. 13 ff.

‡Hayami and Ruttan use an analogous typology in discussing transfers of agricultural technology. They distinguish transfers of material, design, and capability, which as they define them imply the attainment of production, investment, and innovation capability. Yujiro Hayami and Vernon W. Ruttan, *Agricultural Development: An International Perspective* (Baltimore: Johns Hopkins Press, 1971), pp. 175 ff.

operating productive facilities, the second is for expanding capacity and establishing new productive facilities, and the third is for developing technologies. Proficiency in production capability is reflected in technical or x-) efficiency and in the ability to adapt operations to changing market circumstances. Proficiency in investment capability is reflected in project costs and in the ability to tailor project designs to suit the special circumstances of the investment. Proficiency in innovation capability is reflected in the ability to develop technologies that are less costly and more effective.

The boundaries separating these broad areas of technological capability recognize differences that originate in the specialization of activities applying technological knowledge. Table 6.1 provides a list of the principal activities associated with production and investment capability. Much technological knowledge is tacit – it resides as much in minds as in manuals. That is why experience is critical for becoming proficient in each of these activities. But the scope for transferring capability gained through experience in one activity to other activities is limited.[2] For example, the understanding required for repairing and maintaining physical capital has very little in common with that required for optimizing the scheduling of production. For another example, basic engineering requires highly specialized knowledge of the relevant core processes, while detailed engineering activities, such as designing architectural and civil works, require other specialized knowledge.

Moreover, the experience gained in operating, say, a steel mill has little to do with (in the sense that it is not sufficient for) establishing a new steel mill. The differences between the activities involved and the specialized knowledge used are simply too great. In fact, many elements of the experience gained in the operation of one steel mill may not transfer to the operation of another using a somewhat different technology. But the transferability varies across industries. For example, experience operating textile factories may suffice for the establishment of similar factories since textile machinery is available in standardized designs.

Innovation capability pertains to the activities of conceiving and implementing changes in the technical and transactional elements of technology. Included are all the activities spanning invention to innovation, ranging from radical (major) new departures to

TABLE 6.1
Elements of Production and Investment Capability

Production Capability

Production management – to oversee the operation of established facilities

Production engineering – to provide the information required to optimize the operation of established facilities, including:

- Raw material control – to sort and grade inputs, seek improved inputs
- Production scheduling – to coordinate production processes across products and facilities
- Quality control – to monitor conformance with product standards and to upgrade them
- Trouble-shooting – to overcome problems encountered in the course of operation
- Adaptations of processes and products – to respond to changing circumstances and to increase productivity

Repair and maintenance of physical capital – according to regular schedule or when needed

Marketing – to find and develop uses for possible outputs and to channel outputs to markets

Investment Capability

Manpower training – to impart skills and abilities of all kinds

Preinvestment feasibility studies – to identify possible projects and to ascertain prospects for viability under alternative design concepts

Project execution – to establish or expand facilities, including:

- Project management – to organize and oversee the activities involved in project execution
- Project engineering – to provide the information needed to make technology operational in a particular setting, including:
  - Detailed studies – to make tentative choices among design alternatives
  - Basic engineering – to supply the core technology in terms of process flows, material and energy balances, specifications of principal equipment, plant layout
  - Detailed engineering – to supply the peripheral technology in terms of complete specifications for all physical capital, architectural and engineering plans, construction and equipment installation specifications
- Procurement – to choose, coordinate, and supervise hardware suppliers and construction contractors

continued

Table 6.1, Continued

Embodiment in physical capital – to accomplish site preparation, construction, plant erection, manufacture of machinery and equipment
Start-up of operations – to attain predetermined norms

Notes: The elements listed under production capability refer to the operation of manufacturing plants, but similar activities pertain to the operation of other types of productive facilities as well. In turn, our use of "production engineering" departs from conventional usage in that we use the term far more broadly to include all of the engineering activities related to the operation of existing facilities. In our usage, the term encompasses "product design" and "manufacturing engineering" as these terms are generally used in reference to industrial production. See the entries under these headings in the *McGraw-Hill Encyclopedia of Science and Technology* (New York: McGraw-Hill Book Company, 1977).

Source: Larry E. Westphal and others, "Exports of Capital Goods and Related Services from the Republic of Korea," World Bank Staff Working Paper, no. 629 (Washington, D.C.: The World Bank, 1984).

incremental (minor) improvements in existing technology.* Basic research has no direct relationship with innovation capability because its objective is to gain knowledge for its own sake. But there is an obvious and important indirect relationship because basic research feeds into applied research and development, either as the source of new knowledge or as the locus of human and institutional capital accumulation. Innovation capability is directly related to applied research and development, the objectives of which are to obtain knowledge with specific commercial applications and to translate it into concrete operational form. It needs to be recognized, however, that some development activities, such as the creation of pilot plants, also involve production and investment capabilities.

As the examples of textiles and pilot plants suggest, the boundaries that separate the three broad areas of capability – production, investment, and innovation – are fuzzy, particularly in adaptation. In industrializing countries as elsewhere, the successful

*It can be argued that the introduction of existing technology into a new environment requires capabilities that are not unlike some of those involved in innovation, as it is commonly defined. But in our framework this form of "local innovation" entails investment capability rather than innovation capability.

use of technology in production is known to require continuous adaptation involving the activities associated with production capability (see Table 6.1).* Adaptation is important because the optimum utilization of technology is rarely achieved at the outset. Not only is the scope for optimizing within given circumstances almost limitless, but circumstances are constantly changing in input and output markets.[3]

Adaptations are not costless – they require the allocation of resources to purposive technological efforts. The changes are typically implemented through trial-and-error testing of modifications that is like applied research, even though it often does not involve inputs from specialized agents of R & D, either within the firm or outside it. Similarly, adaptation often involves elements of investment capability. This is true because adaptations are frequently embodied in minor changes or additions to physical capital. In some cases these adaptations take place through a process that encompasses all aspects of investment capability, even though the amount invested is small.

More generally, experience-based technological efforts to adapt production technology can provide part of the understanding necessary to carry out some – but not all – of the activities involved in investment and innovation. For example, plant engineers may acquire some capability in plant design and spare parts production from experience in breaking bottlenecks, maintaining equipment, and solving production problems. But it is unlikely that they will acquire a capability in basic plant design, capital goods manufacture, or the creation of radically new technologies. The background and experience needed to carry out many of these activities is different, and the relevant capabilities tend to be developed in specialized entities such as process engineering firms, capital goods producers, and technological research institutes. In turn, the accumulation of local experience in carrying out investments can provide some of the understanding that is relevant to innovation – for example,

*See Dahlman and Westphal, "Technological Effort in Industrial Development." They conclude from the available evidence that "the cumulative sequence of (minor, incremental) technological changes following the initiation of a (successful) new activity (in an industrializing country) may have a greater impact on the productivity of employed resources than that produced by its initial establishment" (pp. 113-17).

knowledge of basic engineering or the ability to embody technology in capital goods. But innovation often involves highly specialized technical knowledge that goes beyond what is needed in relation to investment.

To summarize, the boundaries between areas, though fuzzy, show that specific investments are required to assimilate distinct capabilities. These investments often involve experience in adapting existing technology. Moreover, the differentiating characteristics of the various technological capabilities are loosely reflected in the institutions that organize and coordinate economic activity.

But patterns of specialization and exchange shift both within and among entities, many of which are newly evolved, as existing capabilities are strengthened and additional capabilities are added. The changes are generally, but not always, in the direction of increasing specialization. Where warranted by – among other determinants – the extent of the market, some activities become the domain of specialized units within individual entities or of specialized entities. The latter include a wide variety of firms that provide various marketing, consulting, and engineering services; firms that engage in the manufacture and construction of physical capital; and institutions that perform educational and training services. Producing firms necessarily possess some production capability, aspects of which may be concentrated in differentiated marketing, engineering, and other departments. Nonetheless, such firms may often rely on outside entities for services related to marketing, quality control, process optimization, plant and equipment maintenance, and the like. Moreover, producing firms almost always rely on outside entities for at least some aspects of investment capability: feasibility studies, architectural and specialized engineering designs, and fabrication of physical capital, for example. In turn, specialized innovation capability is found in R & D departments and in separate R & D establishments.

Shifts in patterns of specialization and exchange are important because they are the way that the transactional elements of technology change, and rarely can the full promise of technical changes be realized without changes in these elements.[4] Such institutional change is accomplished through investments that are embodied in the form of organizational structures, codified knowledge and procedures, and customs that govern behavior within and among entities. Thus there is good reason to distinguish institutional from human and physical capital and to say that technological capability resides in both human and institutional capital.

Relationship of Technological Capabilities to Trade in the Elements of Technology

The specialization and exchange in modern economies are such that few entities are wholly integrated systems with all the technological capabilities relevant to their establishment and operation. Instead, technological capabilities are deployed across entities through market transactions in the elements of technology. These transactions take many forms involving goods, services, and information. And they take place between countries as well as within them. Historically, transactions between countries have increased tremendously in volume and diversity as part of the phenomenon of growing global economic interdependence.

For analytical purposes it helps to distinguish several broadly defined elements present – either singly or in combination – in international trade in the elements of technology.

- *Technological knowledge*: the information about physical processes and social arrangements that underlies and is given operational expression in technology.
- *Technical services*: the activities of translating technological knowledge into the detailed information required to establish or operate a productive facility in a specific set of circumstances.
- *Embodiment activity*: the activities of forming physical capital in accord with given and complete design specifications.*
- *Training services*: the activities of imparting the skills and abilities that are employed in economic activity.
- *Management services*: the activities of organizing and managing the operation of productive facilities, the implementation of investment projects, and the development of product and process innovations.
- *Marketing services*: the activities of matching the capacity of productive facilities to existing and latent market demands.

---

*Reasons for separately distinguishing embodiment activity are given below.

The first of these elements is technological knowledge in whatever form it is available without processing or translation to tailor it to (a different set of) specific circumstances. Knowledge in this form is often obtained without payment – that is, without a market transaction explicitly meant to secure it. Most of the other elements are sometimes also obtained without payment, but much less frequently. What distinguishes these is that they consist of activities that reflect the application of technological knowledge in establishing and using productive facilities. Most transactions involving the elements of technology combine technological knowledge with one or more such activities. These activities are the same ones as were discussed in relation to the various technological capabilities; the elements listed above simply cluster them in a different way. The importance of distinguishing the activities from the technological knowledge that underlies them rests in the need for a clear framework to analyze comparative advantage in relation to trade in the elements of technology.

Trade in the elements of technology takes many transactional modes that can serve numerous objectives. Licensing, subcontracting, technical agreements, management contracts, marketing arrangements, turnkey project contracts, direct foreign investment, and trade in capital goods are only some of the forms. Part of this trade simply provides complementary services without any real flow of technology.* Marketing services provided under international subcontracting are a case in point, though they are often combined with technical services that yield an inflow of technology. Several forms of this trade involve packages of more than one element in addition to technological knowledge. Direct foreign investment and turnkey project contracts are obvious examples. In turn, while all forms can play an important role in developing local capabilities, this role is not assured for any of them. Even training activities, which have the ostensible objective of imparting technological capabilities, do not always achieve their goals. Moreover, to the extent that imported elements are not readily available from local sources, imports are more obviously meant – at least in their initial effect – to substitute for or complement local capabilities.

---

*In this respect such trade might preferably be called trade "involving" – rather than "in" – the elements of technology, but we prefer the single label that better describes most of the trade we are concerned with.

Trade in the elements of technology makes it possible to develop most industries through any one of many possible combinations of local and foreign capabilities. At one extreme is the autarkic creation (or re-creation) of technology by locally providing all of the requisite elements of technology through accumulating the relevant technological capabilities. This extreme is likely to prove both very costly and very time-consuming even if use is made of freely available foreign technological knowledge. It nevertheless guarantees the acquisition of at least rudimentary proficiency in the relevant technological capabilities. At the other extreme, an industry can be established and operated on the basis of foreign capabilities with no local accumulation of capability. This sometimes happens, for example, with direct foreign investment in an enclave where indigenous involvement is limited to an unskilled labor force. This extreme may prove an effective way of generating employment and earning foreign exchange in the short to medium term, but it does not contribute to technological development. Between these extremes are many possibilities, some of which are illustrated in the following sections on South Korea's experience.

Although trade in the elements of technology can play an important role in developing local capabilities, the relationship is not a simple one. Trade in these elements transfers the elements, not the capabilities to provide them, certainly not as a direct (or immediate) result of their being imported. To illustrate: Turnkey projects are a frequent vehicle for establishing new industries in developing countries. These projects package all the elements required to establish a production facility and train people in its operation. Rarely do they transfer any of the investment capabilities. Instead, they are intended to transfer production capability.

The relationship is further complicated by the fact that capabilities that are ostensibly intended to be acquired are sometimes not acquired. This is true because technological development can occur only as the result of purposive effort to assimilate technology. There is abundant and compelling evidence on this point from numerous instances of unsuccessful attempts. In turn, capabilities that are typically not expected to be acquired, given the nature of the transaction, can be acquired as the result of purposive effort. Examples are given in the following sections on South Korea.

It is impossible to discuss technological development meaningfully without reference to the objectives sought. Trade in the elements of technology is an important consideration in assessing these objectives. Much technological development is import substitution to replace foreign capability with indigenous capability in activities related to local production and investment. The benefits of this technological development can – and often do – extend beyond simple import substitution to include the ability to adapt technology. Moreover, such development can increase exports, including exports of the elements of technology, as it has in all semi-industrial economies. And there are possible externalities for technological development in other areas to be considered as well. But the accumulation of human and institutional capital in specific technological capabilities also has costs, which include the investment of time and resources needed to accumulate specific capabilities. In addition, as initial experience is being gained, there may be higher costs and greater risks in using local capabilities.

Technological development is not to be seen as having the objective of progressive import substitution for *all* of the elements of technology. Costs and benefits cannot be evaluated without taking account of other opportunities for technological development that are necessarily forgone (or postponed) by dedicating scarce high-level manpower to one area. Particularly in the face of continual global technological development, efficiency in local technological development implies continued imports of many elements of technology, though the pattern of imports shifts as indigenous capabilities are developed.

To conclude: trade in the elements of technology is a critical dimension of any strategy for technological development, which in turn is an important correlate of industrial strategy. Industrial strategies are often discussed simply in terms of the sectoral composition of industry and the market orientation of industrial activity vis-à-vis foreign trade. One of our purposes in this chapter is to demonstrate that technological strategy can fruitfully be discussed in the same terms as industrial strategy, in terms of the sequence of choosing the capabilities to be acquired. As should be apparent from the foregoing discussion, sequencing is not simply a matter of the particular industries in which technological capabilities are acquired. It involves broadening – in part by adding capabilities to accomplish new technological activities in established industries. It involves deepening – by achieving greater proficiency and increased differentiation of existing capabilities. And it involves increasing discrimination in the selection of elements of technology that can be efficiently supplied through local development. In turn, technological capabilities can be employed not only in simple import substitution but also in adaptation and in exporting and can be the base from which additional capabilities are acquired. The following sections discuss South Korea's industrial and related technological development from this perspective.

THE REPUBLIC OF KOREA'S INDUSTRIAL AND TECHNOLOGICAL DEVELOPMENT

South Korea's rapid sustained industrialization dates from the mid-1960s. Most of our discussion is correspondingly focused on the second half of the 1960s and the 1970s. But the heritage from before 1965 needs to be kept in mind when one contemplates how the South Koreans were able to make such effective use of modern technology after 1965. Thus we begin with a brief historical sketch, then turn to the past two decades and examine South Korea's technological development, and conclude the section with a discussion of the acquisition of technological capability.

Historical Perspective

Modern industrialization in Korea began in the colonial period, when the Japanese government managed the peninsula's economy as an integral part of its empire. Manufacturing growth during the colonial period was rapid and extensive, but it depended heavily on the Japanese. Nonetheless, Koreans apparently acquired, mostly on the job, substantial knowledge about how to operate modern industries. There is not universal agreement, however, on just how much human capital was built up during the colonial period. Suh emphasizes the "imposed" enclave nature of industrialization during the colonial period and concludes that the colonial bequest of human capital was negligible.[5] But Mason and others emphasize the "'demonstration effect' of exposure to modern technology and forms of organization" and conclude that the colonial bequest was considerable.[6] We are not sure just how much South Korea's subsequent industrialization owes to the colonial period. All we know for certain is that its colonial experience was unlike that of most other developing countries in that it involved considerable industrialization.

The Korean economy suffered tremendous disruption at the end of World War II when its ties to Japan were severed and the peninsula was divided into two political entities.[7] In what now is the Republic of Korea, or South Korea, manufacturing production in 1945 was substantially less than one-fifth of its level in 1940. But in the light of circumstances at the time, what is really remarkable is that the South Koreans were able, with relatively little foreign managerial or technical assistance, to operate nearly half the manufacturing plants that existed in 1944: there was no sector in which they were unable to produce at least something. By 1948, with assistance from the United States – access to raw materials, replacement parts, and technical help – the South Koreans were operating facilities to produce a wide variety of manufactured goods, including shoes, textiles, rubber tires, basic steel shapes, and such engineering products as pumps, bicycles, tin cans, and ball bearings.

Because of the Korean War, adjustments to dislocation continued to dominate economic activity until the mid-1950s. A respectable but not outstanding rate of industrial expansion was achieved during the latter half of the 1950s, with import substitution for light manufactured and nondurable consumer goods playing the major role. A more important development during this period was the tremendous expansion of education, to which U.S. aid contributed.[8] By 1960 the Republic of Korea had achieved universal primary education, nearly universal adult literacy, and rapidly growing enrollment rates at all levels above the primary level. U.S. aid also financed overseas education and training for thousands of South Koreans. In addition, Americans helped the South Korean military learn modern concepts and techniques of management and organization, as well as how to operate and maintain all types of machinery and equipment. For almost all the male labor force, military service, which was universal, seems to have been an important source of skill formation and general experience in an organization having many characteristics of modern industry.

The South Koreans also gained some industrial competence from their relationship with the United States. Important channels for the direct transfer of industrial technology included the inflow of technical advisers and a modest volume of project assistance. The U.S. military was another channel. Its local procurement program afforded producers in a number of sectors with occasions for assisted learning-by-doing to meet exacting product specifications. Among those benefiting from military purchases were construction contractors, plywood producers, and tire makers – all would later become major exporters.

Large-scale U.S. economic assistance continued through the 1960s. Beginning in the late 1950s, however, there was a gradual shift from grant aid to concessional lending. In addition, the South Korean government began to promote inflows of foreign resources of all kinds from increasingly diversified sources. In the early 1960s, as opportunities for easy import substitution rapidly diminished, industrial growth began to falter. A number of attempts were made at policy reform and economic liberalization in the first half of the 1960s, as policy makers came to accept that rapid economic development depended on export-oriented industrialization and a greater effort to mobilize domestic and foreign resources. These attempts culminated in 1964 and 1965, when a number of reforms were successfully implemented.[9]

Industrial Development after 1965[10]

After 1965 the growth of manufactured exports and the rise in domestic demand fueled a rate of industrialization that was much faster than before. The average annual rate of growth of manufacturing production during 1965-81 was twice the 11 percent rate during 1955-65. Underlying the acceleration of growth in manufacturing, the share of exports in gross manufacturing output increased to roughly one-third in 1981 from less than one-tenth in 1965 – it was nil in 1955. During 1965-81 the ratio of exports to gross national product (GNP) more than quadrupled, and the share of GNP originating in the manufacturing sector more than doubled. Manufacturing exports became increasingly diversified, so that by 1981 South Korea was a major exporter of steel, footwear, machinery and transport equipment, and various manufactures of metal and nonmetallic minerals – in addition to textiles, clothing, and plywood, which had led the initial growth of exports.

There was rapid structural change in manufacturing after 1965. Though the share of textiles, apparel, and leather in manufacturing value added (at 1975 prices) was the same (18 percent) in 1966 and in 1981, the share of food, beverages, and tobacco declined from 37 to 20 percent. Roughly offsetting this decline were increases in the shares of chemical products, including petroleum refining (from 15 to 23 percent), basic metals (from 5 to 10 percent), and machinery and transport equipment (from 13 to 20 percent). South Korea's chemical industry had its start in the early 1960s when large-scale petroleum refining and fertilizer production were introduced. The first petrochemicals complex came on line around 1970. The most notable single event in the development of the basic metals industry was the construction of South Korea's integrated steel mill in the early 1970s. The initial growth of the machinery sector was centered in electrical appliances and electronics. Production of electrical machinery grew far more rapidly than that of nonelectrical machinery throughout the period. In transport equipment the most remarkable achievement has been in shipbuilding, following the establishment of one of the world's largest shipyards in the early 1970s.

Until the mid-1970s the government's strategy largely operated on the accumulation of production capability. Indeed, before the shift in strategy, government policy discriminated against domestic investment capability, most notably by favoring imported capital goods. This policy bias was reversed in the early 1970s, at about the time that the *Heavy and Chemical Industry Plan*, a long-term plan covering the decade to 1981, was published. Among other things, this plan called for the rapid buildup of capacity to manufacture capital goods, particularly the fabricated structural elements and heavy equipment used in industrial plants producing basic intermediate goods and in power generation and transmission and other social overhead facilities.

As will be discussed later, exports of capital goods and related services – in other words, exports of elements of technology – burgeoned after the mid-1970s. But the manner in which they grew was not foreseen by the plan. Indeed, the plan was heavily focused on import substitution, reflecting the government's intention not to increase the economy's dependence on export activity. But very soon thereafter the government abandoned its intention. This, together with gradual recognition of the extent of its overambitiousness, led to several revisions of the plan. In the light of the country's emerging exports of elements of technology, successive revisions became increasingly focused on export activity.

It was also during the mid-1970s that the government began to give serious priority to technological development,* and export activity likewise became an integral part of its efforts to promote the acquisition of technological capability more generally. Separate legislative acts to promote the technological development of producing firms and to encourage the development of engineering services firms included the provision of incentives for exports of the elements of technology. These measures were supplemented by others designed to foster the education and training of qualified personnel in various technical fields and to establish an infrastructure of scientific and technological institutes designed to serve industry.

We turn now to a brief examination of South Korea's reliance on formal technology transfers, starting with the role of direct foreign investment (DFI). Such investment has played only a minor part in the country's industrialization. There was no DFI between 1945 and 1960, when the first legislation controlling nongrant inflows of foreign capital was promulgated. The first instance of DFI in the period after World War II was in 1962. By the end of 1981, the government had approved 693 instances (or cases) of DFI in manufacturing. The cumulative gross inflow amounted to roughly U.S. $1.25 billion. This figure may be compared with cumulative total gross investment in manufacturing over the same period, U.S. $22.7 billion, to show the relative magnitude of DFI.†

A big part of DFI was approved on condition that it involve exports. Nonetheless, it made very little contribution to the expansion of exports during 1962-71 – that is, during the period when export expansion replaced import substitution as the primary engine of industrial development. Indeed, South Korea's first free-trade zone explicitly designed to attract foreign participation in exports

---

*We know that the Ministry of Science and Technology and the Korean Institute of Science and Technology were created in the mid-1960s and were active before the mid-1970s. But we judge that the real shift in priorities came well after their establishment.

†Notwithstanding the relative unimportance of DFI, other foreign capital inflows have been very important as a source of the investment finance needed to achieve rapid industrialization. For example, though the share of capital inflows in total investment has fallen steadily over time, it was 20 percent even as late as 1972-76.

was not established until 1970.* A disproportionate inflow of DFI occurred in 1972-76 (see Table 6.2), much of it oriented toward exporting. DFI intended for offshore production became relatively less important after 1976.

The composition of DFI during 1962-81 is compared with that of value added in Table 6.2. For comparability, values for both are cumulative over the 20 years. The principal sectors for DFI included textiles and apparel, chemicals (synthetic fibers and resins plus petroleum refining and other chemicals), and electrical and nonelectrical machinery. Most DFI in textiles and apparel and a large share of it in electronics were related to production for export and played almost no role in technological development. In the former sector, much of the inflow consisted of the relocation of small plants from Japan to take advantage of low wages. In the latter sector, much was for offshore assembly with very little spin-off to local producers. In contrast, DFI has been a particularly important vehicle for technological development in the establishment of much of the chemicals sector and, more recently, of major elements of the electrical and nonelectrical machinery sectors. DFI has also contributed to technological development in the basic metals sector, but there is no foreign equity in the integrated steel mill.

Formal flows of disembodied technology via other modes also appear to have been rather modest. The cumulative value of technical assistance in manufacturing from bilateral and multilateral sources during 1962-81 was well under U.S. $100 million, as was the cumulative value of technical consultancy by private parties. The latter was heavily concentrated in the chemical and machinery (both electrical and nonelectrical) sectors. Table 6.2 also shows the composition of licensed technology imports. The total number of approved manufacturing technology imports during 1962-81 was 1,840; royalty payments over the same period totaled U.S. $565 million. The volume of licensed imports was rather modest until the mid-1970s. Thereafter the increased reliance on licensing can be explained by the accelerated development of the technologically more advanced industries in recent years. Licensing has been an important source of technology transfer in much the same industries as DFI – chemicals, basic metals, and machinery.

*Investments in it and a second zone established later accounted for roughly 10 percent of the cumulative inflow of DFI by the late 1970s.

International comparisons give a particularly fruitful perspective on the Republic of Korea's pattern of reliance on different modes of technology transfer. Table 6.3 compares South Korea with four other semi-industrial economies – Argentina, Brazil, India, and Mexico – using the best information that we have been able to obtain. Additional cross-country information cited by Westphal and others supports the indication given by the first two blocks of comparative data that DFI and disembodied technology inflows via commercial channels have by no means been relatively large in South Korea.[11] In turn, the third block of data confirms what knowledge of the economy would lead one to suspect, namely that the country's reliance on imported capital goods has, in contrast, been relatively large. Imports of capital goods were more than 20 percent of the value of investment in South Korea throughout the 1970s. The closest country of the other four in this respect was Mexico, with ratios of 11-14 percent during the 1970s.

South Korea's dependence on imported capital goods should be seen as a result of specialization within the capital goods sector and of the demands of a rapidly growing and diversifying industrial sector, rather than as the result of failure to develop a capital goods sector. While it is true that the growth of nonelectrical machinery production failed to keep pace with the growth of industrial output, capital goods production nonetheless grew during 1962-81. The capital goods sector dates back to the colonial period. Over time, all the important metalworking processes, such as casting and machining, were assimilated by South Korean firms and used in copying many types of imported equipment, with the designs subsequently modified on the basis of experience to make them more appropriate to local circumstances. But the capability to design and produce capital goods was oriented toward the more labor-intensive segments of those industries that had a relatively long history in the country. Most export industries used imported equipment extensively, as did most new industries established under government incentives.

Insight about how the South Koreans were able effectively to assimilate technology comes from the comparative data on human capital formation in Table 6.3. What stands out about the educational pattern are the high proportion of postsecondary students abroad, the high secondary enrollment rate, and the high percentage of engineering students among postsecondary students. Even more remarkable is the rapid growth in numbers of scientists and engineers,

TABLE 6.2
Proprietary Transfers of Technology by Sector and Plan Period

	Foreign investment		*Licensed technology*		*Production*
	Approvals granted	*Amount invested*	*Approvals granted*	*Royalty payments*	*Value added*
Percentage distribution of cumulative values by sector, 1962-81					
Food, beverages, and tobacco	2.6	3.6	2.5	1.1	24.0
Textiles, apparel, and leather	9.7	7.2	1.6	1.1	19.0
Pulp and paper products	0.4	0.1	0.8	1.5	2.5
Pharmaceuticals	2.0	1.6	2.9	0.4	*
Synthetic fibers and resins	14.4	30.9	2.7	4.5	2.8
Petroleum refining and other chemicals	1.1	8.2	18.7	36.8	20.2
Cement and ceramic products	3.2	1.6	3.0	2.4	5.1
Basic metals	9.1	6.7	9.7	11.4	7.1
Nonelectrical machinery	17.2	8.9	31.6	21.3	2.3
Electrical machinery	24.8	22.4	19.5	12.3	9.3
Transport equipment	1.0	5.0	3.0	3.3	5.3
Other manufacturing	14.4	3.9	4.0	3.7	2.3
Total manufacturing	100.0	100.0	100.0	100.0	100.0

TABLE 6.2, Continued
Proprietary Transfers of Technology by Sector and Plan Period

	Foreign investment		*Licensed technology*		*Production*
	Approvals granted	*Amount invested*	*Approvals granted*	*Royalty payments*	*Value added*
Percentage distribution by plan period					
1962-66	2.0	2.5	1.6	0.1	4.4
1967-71	20.1	6.7	14.2	2.7	11.5
1972-76	56.4	47.3	22.5	18.4	28.3
1977-81	21.5	43.5	61.8	78.8	55.8
Total 1962-81	100.0	100.0	100.0	100.0	100.0
	Total cases	Total value	Total cases	Total value	Total value
Aggregate cumulative total	693	1,249	1,840	565	156,351

*Value is included in other chemicals

Notes: Values are in millions of U.S. dollars: at current prices for amounts invested and paid; and at constant 1975 prices for value added. Totals may not reconcile due to rounding.

Sources: Information on transfers from Korea Ministry of Finance; that on value added from Bank of Korea, *National Income in Korea*, various years.

TABLE 6.3
Indicators of Technology Inflows, Human Capital, and R & D for Five Semi-industrial Economies

Item	*Year or period*	*Argentina*	*Brazil*	*India*	*Korea*	*Mexico*
Stock of direct foreign investment	1967	10.4	4.0	3.0	1.7	7.3
as a percentage of gross domestic product	1977-79	4.7	6.4	2.1	3.2	5.6
Payments for disembodied technology	1970-71	–	0.20	–	0.04	–
as a percentage of gross national product	1977-79	–	0.33	–	0.17	0.23
Imports of capital goods	1965	5.3	4.6	10.3	13.0	14.5
as a percentage of gross domestic investment	1977-79	8.6	8.4	5.6	27.2	11.8
Postsecondary students abroad	1970	1.0	1.0	1.0	2.0	1.0
as a percentage of all postsecondary students	1975-77	0.3	0.7	0.3	1.7	1.0
Secondary students as a percentage	1965	–	–	29.0	29.0	17.0
of secondary age population	1978	46.0	17.0	30.0	68.0	37.0
Postsecondary students as a percentage	1965	–	–	4.0	5.0	3.0
of eligible postsecondary age population	1978	18.0	10.0	9.0	9.0	9.0
Engineering students as a percentage of total postsecondary age population	1978	14.0	12.0	–	26.0	14.0
Scientists and engineers in thousands	Late 1960s	12.8	5.6	1.9	6.9	6.6
per million of population	Late 1970s	16.5	5.9	3.0	22.0	6.9
Scientists and engineers in R & D	1974	323	75	58	–	101
per million of population	1976	311	–	46	325	–
	1978	313	208	–	398	–

TABLE 6.3, Continued
Indicators of Technology Inflows, Human Capital, and R & D for Five Semi-industrial Economies

Item	*Year or period*	*Argentina*	*Brazil*	*India*	*Korea*	*Mexico*
R & D expenditures as a percentage of gross national product	1973	0.3	0.4	0.4	0.3	0.2
	1978	0.4	0.6	0.6	0.7	–

– Not available.

Source: Project files summarizing data from diverse international sources, World Bank Research Project Ref. No. 672-48. Details are available on request from the authors.

such that by the late 1970s South Korea apparently had by far the highest percentage of scientists and engineers in the population of the five countries. It likewise appears to have had proportionately more scientists and engineers engaged in R & D and to have spent proportionately more on R & D. The well-known problems of measurement and interpretation of R & D statistics dictate the need for caution not to read too much into these last comparisons. We nonetheless believe that they are indicative of Korea's relative ability to undertake technological efforts related to assimilation and adaptation.

The Acquisition of Technological Capability

DFI and the modes of disembodied technology imports discussed above are formal means of technology transfer, as is the import of capital goods, which is typically accompanied by disembodied technology in the form of manuals and training. There also are other, informal modes of transfer that span a wide range of possibilities. Evidence about their importance is difficult to obtain, but there is information. Some of it comes from a survey of 113 exporting firms by Pursell and Rhee in 1976.[12] The sample was meant to be representative of all exporting firms in the Republic of Korea.

The firms were asked about the sources of the basic production or process technologies they then used. Domestic sources were considered to be important slightly more often than were foreign sources. For domestic and foreign sources taken jointly, the sources of technology most frequently cited were buyers of output and suppliers of equipment or materials. Next most important were employees with previous experience working in firms overseas – many as a result of training under turnkey and similar arrangements – and in South Korean establishments. Indeed, the transfer of labor among firms was more important than contacts with suppliers alone or with buyers alone.*

---

*The importance of labor transfer as a source of technology reflects high labor mobility. Depending on the industry, between 33 and 51 percent of the production workers recruited to individual firms in 1975 had previous experience in the job assigned to them.

Formal mechanisms of licensing and technical assistance, of only modest importance overall, were the most important modes for transferring technologies from abroad. Even so, they were considered to be important only one-third as often as foreign sources were indicated. In turn, foreign buyers contributed informal transfers of technology, frequently as a result of periodic visits to inspect production facilities or of ongoing programs to control and improve quality. Through such things as suggesting changes in individual elements of the production process and improvements in the organization of production in the plant and in management techniques more generally, buyers helped many exporters achieve greater efficiency and lower costs. There can be no doubt that the transfer of know-how from export buyers has been a contributor to minor process innovations of the sort that sequentially lead to gradual improvements, the cumulative effect of which can be great.

For many industries it is important to distinguish between the mastery of production processes and the ability to design products that either conform to the structure of or anticipate changes in demand. South Korean exporters, almost across the board, relied heavily on foreign buyers for product design technology, far more so than for process technology. Foreign buyers contributed to product innovation through the influence they exercised on the characteristics of exported products. Nearly three-quarters of the sampled firms stated that they either modified the characteristics of their product to accommodate buyers' requests or produced in direct accord with buyers' specifications. Nonetheless, the majority of firms produced only some of their exports directly to buyers' specifications. Those most often influenced were product design and styling, followed by packaging, basic technical specifications, and minor technical specifications.

It is not surprising that South Korean exporters have relied extensively on foreign sources for product innovation. This is inevitable when technology is transferred to start new lines of production that serve export markets from their inception – as in much of local shipbuilding production, for example. But reliance can also develop if production has first been established to serve the local market, with exports following later, as often occurred in South Korean experience. Here mastery of technology is in the first instance often confined to achieving rudimentary standards of product design. These may suffice to gain entry into export markets, but continued growth of exports sooner or later requires that product standards be upgraded. Moreover, successful penetration of export markets frequently requires that product specifications be tailored to the different demands of individual markets. Until some experience has been gained in producing to meet differentiated demands, it undoubtedly is most cost-effective, and it may even be necessary, to rely on export buyers for product design technology. Not to be neglected in this regard is that production for export provides a potent means of acquiring product design technology through learning-by-doing, which spills over to product development in local markets as well.

The results of the survey of exporters clearly indicate that the acquisition of technological capability by South Korean industry in basic production processes had progressed further than in product design, at least in relation to product standards in developed country export markets. In addition, given the high frequency with which domestic sources of process technology were said to be important, the results attest to considerable local mastery of basic production technology. Much of what was considered by the respondents to have come from domestic sources consisted of technology originally developed overseas, subsequently transferred or brought to South Korea, and then effectively assimilated and sometimes adapted by indigenous industry. Some of this technology, particularly in the traditional export sectors, was part of the inheritance from the colonial past.

The basic production technology for nonsynthetic textile yarn and fabric is an obvious case: among today's leading textile exporters are several that were established before independence, and some senior managers and technicians who gained their initial experience during the colonial period were still active in the 1970s. Plywood – also an important export, particularly during the 1960s – offers another example: the first plant was constructed in 1935. Plywood is equally an example of an industry that benefited from technical assistance provided under the U.S. military's program of local procurement during the 1950s. Nonetheless, when queried about the sources of technologies in use today, producers of both textiles and plywood overwhelmingly indicated local sources.

The distinction between domestic and foreign sources thus has little to do with where the technology was invented. It has far more to do with the importance of the assimilation and adaptation of technology by local producers and of the diffusion of technology through formal and informal contacts and through labor transfers among domestic firms.[13] Further evidence of the importance of diffusion from domestic sources was found in the sizable number of exporting firms that indicated direct knowledge of diffusion to other firms of technologies they had introduced into the country.

In industries for which process technology is not product specific, mastery has frequently led to the copying of foreign products as a means of enlarging technological capacity. The mechanical engineering industries, among others, afford many examples; such processes as machining and casting, once learned through producing one item, can easily be applied in the production of other items. One case that has been closely studied is textile machinery, particularly semiautomatic looms for weaving fabric.[14] In this as in some other cases South Korean manufacturers have not only been able to produce a capital good that meets world standards, albeit for an older vintage; they have, in addition, adapted the product design to make it more appropriate to local circumstances. (The adapted semiautomatic looms fall between ordinary semiautomatic and fully automatic looms in terms of the labor intensity of the weaving technology embodied.)

In other industries in which technology is more product specific, such as chemicals, mastery of the underlying principles has enabled greater local participation in the technological effort associated with the subsequent establishment of closely allied lines of production. Recognition of the importance of local technological learning also is central to understanding how technologies initially introduced in South Korea only very recently – within the past five to ten years – are now considered, in relation to subsequent undertakings in the same lines of production, to have come from local sources.

Detailed information about South Korea's technological capability in particular industries is scanty. Still less information is available about the processes by which its capability has been acquired. In preparing this chapter, Kim undertook an extensive survey of the documentary information available from the records of various government and financial institutions, the objective being to determine whether "technological histories" for specific industries could be compiled from existing project files. He also surveyed existing histories of subsectors and firms. The results were, to say the least, disappointing. It appears that access to historical records is not publicly available, and intensive firm-level interviews provide the only feasible ways to compile even sketchy technological histories. Though disappointed by this finding, we nonetheless feel that there is considerable value in knowing, with some confidence, that there is no feasible alternative.

Detailed historical evidence helps in understanding the evolution in the direction of South Korea's technological development. The rapid growth and increasing diversification of exports of all kinds give the most compelling evidence of the country's acquisition of technological capability over time. But this evidence is indirect and is therefore apt to be misleading in certain important respects. Thus it is important to know that exports have not depended crucially on DFI or other forms of international subcontracting.[15] The survey of exporters clarifies that industrialization up to the late 1970s was based largely on proficiency in production engineering, more specifically on mastery of production processes as opposed to design adaptation. In turn, the extent and nature of the acquisition of investment capability did not become readily apparent until the late 1970s, when South Korea began to experience rapid growth in its exports of elements of technology. But detailed case histories make it clear that a process of increasing involvement in project execution had become entrenched in local industry much earlier.

Amsden and Kim have obtained histories of several turnkey plant exporters in their ongoing research on South Korea's acquisition of technological capability.[16] Two distinct patterns of technology assimilation are apparent. One of them, the apprentice pattern, is well illustrated in the following examples.

- *A chemical company that produces soda ash and other products.* Its initial soda ash plant was built on a turnkey basis by Japanese firms with Japanese technology. Participation in the plant's construction enabled the firm's engineers to do subsequent expansions without foreign assistance. Experience gained through operating the plant and a subsidiary that produces white cement, together with design capabilities accumulated through expansion projects and the establishment of other related plants, gave the firm turnkey capabilities to export a white cement plant.
- *An independent chemical engineering company with three chemical manufacturing affiliates and one construction subsidiary.* To gain engineering experience and exploit the local market

for chemical detergents, the firm imported foreign technology to establish a nonbiodegradable chemical detergent plant. The plant was of West German design, with basic technology from a U.S. company; most of its equipment was imported. The firm's engineers took part in all stages of project execution and benefited from intensive training by the U.S. company. The technological capability so obtained later enabled the firm to build a biodegradable detergent plant without foreign technical assistance. Sixty percent of the equipment installed in this plant was made in South Korea. Experience in these projects and in numerous other engineering projects provided the capabilities to export an unsaturated polyester resin plant to Saudi Arabia. Except for basic engineering of the core processes, the firm undertook all stages of project execution.

Many other industries either are known or appear to have followed the apprentice pattern. For example, Kim, in his supplementary survey for this chapter, obtained sketchy information sufficient to confirm that this pattern prevailed in cement and synthetic fibers (specifically, nylon and polyester).[17] The cement industry is a particularly interesting case because of the continuation of selective foreign participation in basic and detailed engineering, plant start-up, and assistance to quality control in recent projects. Continued reliance on foreigners seems to be linked to efforts to remain close to the global frontier in process technology.

Under the apprentice pattern, the first plant in the industry was typically built on a turnkey basis with indigenous involvement "limited" to assimilating as much of the production *and* investment capability as was practical. The development of investment capabilities began with the participation of South Korean engineers – often more as observers than as anything else – in the initial project execution and continued through experience gained in production and plant expansion projects. Construction of the second and subsequent plants followed quickly, with local engineers and technicians assuming a rapidly expanding role in project design and execution. Indigenous involvement in project implementation expanded through concerted effort to assimilate the know-how involved in project design and execution. The process was effectively one of highly selective and experience-centered import substitution, in which successively more complicated aspects of investment capability were acquired and put into practice. The result was a growing capability in all elements of investment activity.

The rapid accumulation of investment capability through apprenticeship is well illustrated by the expansion path of South Korea's integrated steel mill, which began operation in 1973. The first phase, a turnkey operation by foreigners, created annual capacity of 1 million tons. By 1981 capacity had been increased to 8.5 million tons in three expansion phases that were increasingly under the direction of South Koreans. The progressive substitution of local for foreign investment capability is indicated by the fall in the cost of foreign project engineering per ton of incremental capacity: U.S. $6.13, U.S. $3.81, U.S. $2.42, and U.S. $0.13 respectively in the successive phases. There was also a progressive increase in the share of locally supplied capital goods. Initially 12 percent, it rose to almost 50 percent in the fourth phase. Increased proficiency in investment is further reflected in the shorter times needed to execute successive expansions, even though each phase involved a larger capacity increment: 38 months for the initial, one-million-ton phase, falling to 27 months for the fourth, three-million-ton phase. The time required from start-up to achieve rated capacity also fell in each successive phase. Successful apprenticeship involved large investments in training both at home and abroad. Through 1983 the company had sent 1,850 engineers and technicians for training overseas, much of it involving on-the-job experience.

The other pattern of technological development that appears often is the imitator pattern. It is illustrated in the histories of two other turnkey plant exporters.

- *A local manufacturer of steel pipe, with a subsidiary that manufactures steel-pipe-making equipment.* After the Korean War the firm developed a primitive steel pipe manufacturing line, using automotive parts and other war surplus equipment, based on observation of an imported steel-pipe-manufacturing line used by a local bicycle manufacturer. It later developed and improved six separate lines for its own use before importing several more-advanced lines from Japan. Engineering capabilities developed over time, and insights from operating the Japanese lines led the firm to develop several modern lines of its own based on them. The firm's first turnkey export (to Kenya) was a renovated line that was about to be scrapped because of economic obsolescence.

A second turnkey export (to Bangladesh) was designed and manufactured by the firm. The third was a more sophisticated line than the one used by the firm itself. The Saudi Arabian importer specified automated galvanizing lines to save labor.

- *The only specialized steel rolling mill manufacturer in South Korea.* Originally a machinery repair shop, the firm developed a simple, inexpensive hot-rolling mill for a client. Basic design ideas came from observing an imported mill used by a local firm. Using experience gained from previous projects and technical information obtained from observing more sophisticated foreign mills operating locally and abroad, the firm developed the capability to manufacture a wide range of rolling mills. It designed and sold 102 mills for domestic use before exporting a turnkey plant to Egypt.

As illustrated by these examples, in the imitator pattern, local firms started with small and rather primitive technologies developed by themselves and gradually upgraded both processes and products through operating experience and using technical information and ideas that came from observing foreign technology. Other capital goods producers known to have followed this pattern include suppliers of machinery for textile weaving and paper manufacturing. The fast pace of South Korea's industrial growth has been an important factor in both this and the apprentice pattern. The short intervals between the construction of successive plants to which it led has greatly facilitated experience-based learning.

One of our interests in compiling technological histories was to get some sense of the relative importance of formal and informal technology transfers. Informal transfers have clearly dominated in the imitator pattern and have for a long time been significant in broadening all exporters' capabilities. A study by Kim indicates that they have also been important in recent innovations by small-scale (less than 100 workers) capital goods producers, many of whom initiated new product lines by imitating foreign equipment – by copying imported models and using information from sales catalogues or from visits to foreign manufacturers.[18] We cannot say in precise quantitative terms how important informal transfers have been more generally. But everything we know from firm interviews and other sources indicates clearly that they have been very important.[19]

EXPORTS IN RELATION TO TECHNOLOGICAL DEVELOPMENT[20]

South Korea first adopted a strategy of export promotion in the expectation that it would accelerate growth by relaxing the foreign exchange constraint and increase efficiency through resource allocation in line with comparative advantage. These expectations were more than fulfilled. Since the strategy's adoption, exports have led the country's economic growth. Exports also led the economy's development in a more fundamental sense: in the establishment of new industries and in the acquisition of added technological capability in existing industries.

In this section we first amplify these interactions between export activity and technological capability in general terms. Next we use South Korea's exports of the elements of technology to illustrate the interactions in specific terms. We then use the information about these exports to draw some conclusions about South Korea's technological capability and its comparative advantage in technological activities.

Exports and Technological Capability

South Korea's initial export success came largely in industries, such as textiles, established during the Japanese colonial period, before 1945. Exceptions, such as wigs, used technologies easily assimilated, given the technological capabilities then existing in closely related areas. Thus the underpinnings of export performance during most of the 1960s can be satisfactorily understood in the conventional paradigm of static comparative advantage. Exports were concentrated in industries in which South Korea either already had or could easily acquire the needed technological capability. Moreover, these industries had factor intensities in line with the country's relative factor endowment. The predominant gains from these exports were the obvious ones, greater capacity use and increased allocative efficiency.

It was in the late 1960s that export activity became important in establishing new industries in which South Korea did not already have technological capability. Two wars and their aftermath had opened a hiatus in the establishment of new industries. This was closed in the early 1960s with the inception of two industries – chemical fertilizer and oil refining – both of which were to serve the domestic market with no expectation of substantial exports. Then, in the late 1960s, some new industries were established to serve both the domestic and export markets.

Several of these new industries were characterized by pronounced economies of scale. Constructing plants at scales sufficient only to meet expected domestic demand would have resulted in production costs well above internationally competitive levels. Thus exports were used to gain the economies of scale needed to realize the potential comparative advantage that South Korea had in such industries. Here, a notable example is the integrated manufacture of basic steel products.

Some other new industries had negligible domestic sales at their inception. A few of these, electronic components being an early example, were created by direct foreign investment or relied on other forms of international subcontracting for technology transfer and market access. The rest, such as large-scale shipbuilding, obtained their technology through licensing and turnkey plant contracts and did not have guaranteed markets to begin with.

Export activity can enlarge technological capability in two ways, by facilitating technology transfer and by stimulating technological effort. Transfers of technology often accompany DFI to establish plants that are designed to produce for export. They may also be an integral part of transactions involving other means of international subcontracting. But there are few such instances in South Korea, since DFI and international subcontracting have not been very important in most exports. Export activity can also lead to transfers after the acquisition of rudimentary technological capability. Though less obvious, these transfers have been very important, as was noted previously.

In turn, though there is little direct evidence about the effect that export activity may have in stimulating technological effort, it undoubtedly enforces and fosters the acquisition of technological capability. Exporting requires the ability to meet product specifications at a competitive price, and the drive to penetrate overseas markets stimulates efforts leading to the gradual upgrading of product quality. These may even be the most important ways in which export activity adds to technological capability. But this cannot be directly inferred from studies showing that export activity has a strong positive effect on factor productivity.[21]

In sum, under South Korea's strategy of export-led industrialization, export activity has been important in exploiting static comparative advantage. It has also been important in dynamically changing comparative advantage through accelerating the broadening and deepening of industrial competence. Export activity made it possible to start new industries much earlier than they could otherwise have been established without sacrificing economies of scale. In turn, for all industries, and for a long time after their inception, export activity added to technological capability, reflected in a wide variety of minor technological changes.* Thus export promotion has been a strategy as much for developing industry as for capitalizing on the industrial competence at each point along the way. This is well illustrated in the recent growth of South Korea's exports of the elements of technology.

Exports of the Elements of Technology

Over the past decade South Korea has experienced impressive growth in its exports of the elements of technology, which might also be called exports of capital goods and related services. These exports comprise all flows that involve the transmission of technological knowledge and the performance of activities that reflect its application in establishing and operating productive systems. The constituent elements were indicated and discussed earlier. Here we need only add some remarks about the character of embodiment activities, which, it may be recalled, are those of forming and maintaining physical capital in accord with given and complete design specifications.

---

*To the degree that improvements in productive efficiency and other forms of technological change derive from experience in production and capacity expansion, export activity must necessarily lead to faster productivity growth if it is associated with greater volumes of production over time. This argument is the basis for Abegglen and Rapp's conclusion that export activity has played a major role in the growth of Japanese industrial productivity. Our argument is somewhat different in alleging that export activity has substantial effects that are typically not associated with production for the domestic market. James C. Abegglen and William V. Rapp, "The Competitive Impact of Japanese Growth," in *Pacific Partnership: United States–Japan Trade*, ed. Jerome B. Cohen (Lexington, Mass.: Lexington Books for the Japan Society, 1972), 19-50.

Some capital goods exports consist simply of the manufacture of machinery and equipment (or parts thereof) and the fabrication of structural elements in accordance with detailed design specifications by the purchaser (or an agent acting on its behalf); that is, they consist simply of embodiment activity. Others of these exports, though made to order, may also include domestic design engineering, the scope of which depends on the completeness of specifications provided by the purchaser. Still others comprise machinery and equipment routinely produced in the country. These last two categories of exports also consist of embodiment activity. And insofar as they do not simply embody foreign designs obtained under license, they incorporate capability in the design of capital goods; that is, they incorporate an implicit element of technical services.

Overseas construction, in turn, often consists simply of embodiment activity, as when labor services are supplied with a complementary flow of management services.* The only analytically important difference between this kind of overseas construction and exports of capital goods made to conform to given, detailed design specifications is in the embodiment activity performed. And even this difference disappears for some kinds of overseas construction. Metalworking, as distinct from construction, is the primary embodiment activity in exports of capital goods. But metalworking can also be an important part of overseas construction, as it is, for example, in the erection of chemical plants.

Like other advanced, semi-industrial economies, the Republic of Korea is a substantial exporter of many of the elements of technology. But among these countries it stands out as having the largest volume of such exports.[22] One reason for this is the government's active encouragement of these exports. Also important is the rise to prominence in the mid-1970s of the South Korean *chaebol* – large, conglomerate business groups centered on diversified trading companies. These firms have been the leading agents in the export of the elements of technology.

Table 6.4 provides summary data on what we will call project-related exports, which include all forms of the exports with which

* Exports of the elements of technology do not include the services of migrant labor when these are not associated with any other elements, nor do they include flows that consist simply of financial capital or intermediate inputs.

TABLE 6.4
Licensed Project-related Exports by Kind and Sector
(Cumulative, through the end of 1981)

(In millions of U.S. dollars)

	Manufacturing	*Social overhead*	*Services*	*Total*
Overseas construction	2,055	41,777	121	43,953
Plant exports	472	2,098	–	2,570
Direct investment	67	36	–	103
Disembodied technology:				
Licensing and technical agreements	139	13	14	166
Consulting services	155	72	79	306
Total	2,886	43,996	214	47,096

– Zero or not separately distinguished.

Notes: See the text for definitions of kinds and caveats about double-counting. Totals may not reconcile because of rounding error.

Source: Westphal, Larry E., Yang W. Rhee, Linsu Kim, and Alice Amsden, "Exports of Capital Goods and Related Services," from the Republic of Korea, World Bank Staff Working Paper no. 629 (Washington, D.C.: World Bank, 1984), Table 2.

we are concerned except some types of capital goods. Five kinds of project-related exports are distinguished in the table: overseas construction, plant exports, direct (overseas or foreign) investment, licensing and technical agreements, and consulting services. The data are for licensed exports and were tabulated using information from the ministries responsible for licensing them.* (Licensing is done as part of the process of administering export incentives and applies to all exports in the country.)

Overseas construction refers to contracts for construction projects in which the contracting South Korean firm provided more than the services of migrant labor. More will be said about the content of these exports below. Some but not all capital goods exports are included under plant exports. In local usage, plant exports include complete productive systems (such as manufacturing plants and social overhead facilities) and individual elements of such systems (such as textile machinery and distribution transformers). The elements differ from other capital goods exports in that, to be designated as plant exports, they must be purchased for installation in specific productive systems. Moreover, to qualify as a plant export, the transaction has to exceed a minimum value (U.S. $100,000 during the period covered) and satisfy minimum local content requirements that differ by the kind of plant export.

The distinctions among the five kinds of project-related exports are not clear-cut. Most important, contracts that packaged several activities may have been licensed under the dominant activity alone, with the licensed value reflecting the value of the entire package. We know only that this was not a uniform practice. In addition, the reporting of each type of export by a different administrative agency may result in some double-counting. Where direct investment is involved, there is double-counting insofar as the local equity contribution was not in cash but in kind, as frequently was the case. In short, the figures in Table 6.4 are to be taken as indicative rather than definitive. But our impression is that the misclassification and double-counting in these figures is not great enough to make them useless for identifying the central tendencies in Korea's project-related exports.

---

*Data for licensed exports pertain mainly to the value of valid contracts made. Given the lag between contract negotiation and project implementation, the data somewhat overstate the values of actual flows of goods and services.

Overseas Construction

South Korea got into the business of project-related exports through overseas construction, which remains far and away the largest component of these exports. The cumulative value of licensed overseas construction at the end of 1981 was $44 billion, compared with $47 billion for all project-related exports. The composition of contracts for overseas construction exhibits considerable diversity in the size and type of project.

Much of the initial competence in construction activity came from learning-by-doing gained through contracts under U.S. military procurement. Overseas construction began in 1966 and was for some time concentrated in Southeast Asia, where it largely served U.S. military procurement in Guam and Vietnam. The experience gained serving the U.S. military outside South Korea added further capability, specifically in construction work overseas. Thus the stage was set for the dramatic and sustained increase in these exports in 1975, when South Korean contractors began to take part in the Middle East's building boom. Roughly 90 percent of the cumulative value of all contracts has been for work in the Middle East.

Overseas construction appears to consist largely of embodiment activity, though it also includes related organizational and managerial services. Precise information is lacking on the incorporation of technical services related to project design. But the design engineering was South Korean on smaller projects involving buildings – such as schools, warehouses, office structures, apartment blocks, and the like – and simple infrastructure. In turn, the design engineering needed for some large projects, such as those involving large-scale industrial plants, was provided under separate contracts with non-Korean firms.

Plant Exports

South Korea's plant exports began as an adjunct of its overseas construction activity, and further impetus came from the government's promotion of the capital goods sector in the 1970s. Though moving at a much slower pace than the government first hoped, the capital goods sector expanded its capacity quite rapidly, diversifying into new areas. Exports of capital goods in current prices grew at the rate of 60 percent a year during the 1970s, compared with 35 percent for total commodity exports.

Plant exports first reached substantial proportions in 1977. From then onward they may have accounted for as much as one-third of the country's exports of capital goods, with most of the remainder comprising ships and railway vehicles. The cumulative value of the 276 plant exports licensed by the end of 1981 was nearly $2.6 billion, or equal to about 6 percent of the value of licensed overseas construction. Licenses ranged from a U.S. $100,000 contract for power transformers in Indonesia to a $209 million contract for a cement mill in Saudi Arabia.

The provision of a wide variety of manufacturing plants and their elements constituted nearly one-fifth of the value of licensed projects, and more than one-third of the number. The difference in shares by value and by number reflects the small size of manufacturing projects, considerably less than half the average value for social overhead projects. The most important markets for South Korea's manufacturing projects were the Middle East, which accounted for almost 60 percent of the total licensed value, followed by developing countries in Africa and Asia, with roughly 20 percent in each region.

Social overhead projects made up the lion's share of plant exports. They were far more numerous and had an average licensed value of $12.1 million. Very large projects for offshore drilling and coastal facilities were responsible for more than two-thirds of their licensed value. Other projects were related to desalination plants, power generation and transmission facilities, onshore structures, communication facilities, and water treatment plants. Many social overhead projects appear to have been construction related in that they included on-site construction or erection and installation as important elements. The Middle East initially was South Korea's largest market for social overhead projects, but more recently, Organization for Economic Cooperation and Development (OECD) countries have overtaken it. Other developing countries, in Asia and to much less extent in Africa, formed the remainder of the market.

Amsden and Kim have tabulated the elements of 128 plant exports licensed in 1980 and 1981, exports that accounted for nearly half the value of plant exports licensed through 1981.[23] Equipment (including fabricated structural elements) alone was provided in 43 percent of the cases, accounting for 23 percent of the value of the complete sample; equipment and design were provided in 31 percent of the cases, accounting for 35 percent of the sample's value. Fifteen percent of the cases, accounting for 41 percent of the value, were turnkey projects. The remaining 11 percent of the projects, accounting for about 1 percent of the value, combined several elements including equipment but were not full turnkey projects. Both manufacturing and social overhead projects were included in all of these categories.

The absence of design services from some of the manufacturing projects does not mean that the contractor simply provided embodiment services in these cases. On the contrary, most of these projects appear to have supplied South Korean-designed equipment. The absence of design services simply indicates that the designs were standard South Korean designs rather than special ones tailored to the project being served. But circumstances appear to have been different in the social overhead projects: contractors who provided equipment alone probably undertook only the embodiment activities. In turn, design in social overhead projects typically seems to refer to detailed engineering following basic engineering designs provided by the project sponsor. In manufacturing projects, design often appears to have included basic engineering as well.

The driving force behind South Korea's plant exports for social overhead projects was its rapidly advancing comparative advantage in embodiment activity, particularly in large-scale construction projects and capital goods production. The underlying technological capability appears to reside primarily in the organization, management, and execution of construction activity and in the production engineering aspects of metalworking. Additional forces seem to have been at work in the exports of manufacturing projects. Not only did many of these appear to include more sophisticated technical services, such as basic design engineering, but many also appear to have transferred process technologies that incorporated South Korean adaptations. Moreover, several turnkey manufacturing projects called for considerable ability in the organization and management of complex overseas undertakings. In this respect the projects reflected a higher degree of entrepreneurial ability than was probably characteristic of most social overhead projects. Thus the technological capability underlying the manufacturing projects was more extensive in that a much wider range of elements was involved.

Direct Investment

A steady flow of outward investment from South Korea started in about 1967, with a few cases in the preceding years. By the end of 1981 the cumulative licensed outflow of direct investment was U.S. $323 million, of which U.S. $103 million was in manufacturing or social overhead sectors. Thirty-four manufacturing ventures were licensed by the end of 1981. The smallest was for U.S. $25,000 in a printing firm in Japan; the largest was for U.S. $25.7 million in a cement mill in Malaysia. About three-quarters of this investment was in developing countries in Asia; most of the rest was in Africa and the Middle East.

There was a considerable overlap between overseas manufacturing ventures and manufacturing plant exports. Indeed, in some cases an equity position is either known or appears to have been taken to facilitate the export of a plant. But some investments were motivated more by the desire for access to natural resources, albeit in a processed form, than by anything else. Another factor motivating some joint-venture investments, particularly those in the OECD countries, was the desire to gain access to foreign technology through collaboration with foreign companies.

Disembodied Technology Exports

In South Korean usage, licensing and technical agreements are distinguished from technical consulting services. But the dividing line between the two categories is not clear-cut. The only real distinction we know of is that no contracts involving royalty payments over time (as opposed to one-time lump-sum payments) are licensed under consulting services.

No licensing or technical agreements were recorded before 1978, when two contracts were licensed – we do not know their value. Of the 39 contracts worth U.S. $166 million licensed during 1979-81, 36 were with firms in developing countries, mainly in the Middle East. Most licensing and technical agreements provided general technical assistance in production engineering and maintenance, many over several years. The sectors served were mainly in manufacturing, with a rather wide dispersion among these sectors. And some of the contracts were to plants in whose establishment South Korean firms had participated through overseas construction or plant exports.

Exports of consulting services were far more numerous than those of licensing and technical agreements. The cumulative value of the 324 contracts licensed from 1973 through 1984 was U.S. $306 million, with more than three-quarters of the contracts (by number and value) having been licensed in 1979-81. Most comprised technical services accomplished by the dispatch of personnel to oversee construction and plant erection, install machinery and equipment, inspect structures and equipment, troubleshoot, train labor, and the like. The value of licensed contracts was split about evenly between manufacturing and social overhead sectors. Many consulting services, particularly in the social overhead sectors, appear to have been related to construction exports. Firms specializing in engineering appear to have been responsible for most exports of both licensing and technical agreements and consulting services – not only exports related to project execution, but also those involving production engineering. These exports thus reflect the growing capability of South Korea's recently established engineering firms.[24]

Revealed and Dynamic Comparative Advantage

We have discussed five kinds of project-related exports – overseas construction, plant exports, direct investment, licensing and technical agreements, and consulting services. These exports and those of non-project-related capital goods constitute the exports of elements of technology. Overseas construction and plant exports for social overhead projects have predominated. They accounted for more than 95 percent of the cumulative value of licensed project-related exports at the end of 1981 and around half to three-quarters of total exports of capital goods and other elements of technology during 1977-81. The bulk of this export activity appears to have been performed in accordance with detailed specifications provided by the purchaser. We thus conclude that South Korea's revealed comparative advantage in exports of the elements of technology is in project execution, mostly in embodiment activities in the form of construction and metalworking (including erection and fabrication).*

*We are constrained to assess revealed comparative advantage by the composition of Korea's exports. But we doubt that the use of more sophisticated methods would change the conclusion.

These embodiment activities are moderately intensive in human capital. They require reasonably skilled workers and technicians as well as competent engineers, of whom the country has a comparative abundance. But more than this, they require the ability to organize and manage undertakings that are often complex, even in subcontracts. Here South Korean firms appear to have an advantage that permits them, for example, to complete projects in far less time than is considered average or normal. Precise information about this is lacking but anecdotes abound. Moreover, much of the marketing of these exports is done by South Korean firms acting without foreign agents. This is one area where South Korean know-how relating to transactions is second to none.

In most of its exports of the elements of technology South Korea is not exporting newly created technological knowledge. Instead, these exports largely consist of certain project execution activities. In other words, what underlies the country's revealed comparative advantage is its mastery of production engineering in construction and metalworking. It enjoys a cost advantage in these activities due both to its mastery and to its comparatively low wages and salaries – adjusted for skill and productivity differences – for skilled workers, technicians, and managers.

A small part of South Korea's exports nonetheless do appear to transfer idiosyncratic manufacturing technologies created through experience-based adaptive engineering. In turn, technological idiosyncrasies may be an element in the revealed comparative advantage in project execution. That is, the technologies used to carry out these activities might be idiosyncratic. Differences in technology related to organizational modes and procedural methods might, for example, be what underlies the ability of firms to complete projects in record time. This possibility is certainly consistent with our conception of the elements of investment capability.

The technology factor that underlies most of South Korea's exports of the elements of technology is much the same as that which underlies most of its (other) manufactured exports. Insofar as the former exports largely reflect mastery of what might be termed the production engineering aspects of project execution, the rapid growth of both kinds of exports reflects the rapid accumulation of proficiency in production. But the emphasis on proficiency in production should not imply the lack of a design capability. Exports of idiosyncratic manufacturing technology indicate that there is some capability in the design of machinery. And South Korea does export detailed project-engineering services. But these exports are only a small fraction of the total. What is more significant is that South Korea does not appear to possess much capability in basic project engineering.

In discussing the general relationship between exports and technological capability in South Korea, we emphasized two distinct objectives – exploiting the country's existing comparative advantage, and dynamically changing its comparative advantage. There can be little question, especially in the light of the foregoing discussion, that South Korea's exports of elements of technology exploit its existing comparative advantage, in both its endowment of human capital and its mastery of the production engineering aspects of project execution. What, then, of their role in dynamically changing the country's comparative advantage?

Because of their highly specialized nature, many activities of project execution are characterized by extreme economies of scale.[25] South Korean firms could not be internationally competitive in these activities if they served only the domestic market. Export activity has thus made it possible to establish investment capabilities that could not otherwise have been realized without tremendous sacrifice of scale economies. It has also enabled greater continuity over time in the use of capabilities once established, thereby precluding their atrophy owing to infrequent use. Moreover, the accumulation of experience is a critical input in acquiring most of these capabilities.[26] Export activity not only compresses the time for experience to be accumulated; it also affords a wider variety of experience in more diverse circumstances. It can thus be expected to accelerate cost reductions from learning and to deepen existing capabilities.

These benefits appear to be reflected in changes over time in the composition of South Korea's exports of elements of technology toward increasingly more complex and sophisticated activities. Managers of exporting firms have also confirmed, in interviews with Amsden and Kim, that these benefits are realized.[27] But these managers have indicated another benefit as being even more significant: participation in project execution with foreign firms from third countries has been an important vehicle for acquiring additional capabilities and new technologies. These gains have occurred through a process akin to apprenticeship – a process also observed (as previously noted) in South Korean participation in local projects. This broadening of technological capability occurs even in overseas construction. For example, one South Korean firm assimilated a complete system of solar energy technology through its participation with a U.S. firm in a project in Saudi Arabia.

The broadening and deepening of South Korea's industrial competence, particularly its investment capability, appears to be an important motive in the government's promotion of exports of the elements of technology. In the short run, the gains from exploiting existing comparative advantage in these exports are considerable. But in the long run, the gains from further developing the country's technological capability may well be even more considerable.

CONCLUSION

Several things stand out in the Republic of Korea's pattern of technological development. One is the limited extent of reliance on proprietary transfers of technology by means of direct foreign investment and licensing agreements. Among formal transfers of technology, turnkey plants and machinery imports have played by far the greater role. Moreover, in only a few sectors, such as electronics, have exports depended crucially on transactions between related affiliates of multinational corporations or on other forms of international subcontracting. Another prominent feature of South Korean technological development is the apparently tremendous importance of a wide variety of informal transfers that have involved imitation and apprenticeship as well as the use of information obtained in exporting. Also to be included under informal transfers is the expertise that has been obtained as a result of the return of South Koreans from study or work abroad unassociated with formal training, though the importance of this transfer relative to formal training is not known.

Also striking is the selectivity of South Korean technological development and the part played in it by both imports and exports of the elements of technology. South Koreans have acquired a good deal of technological capability, but they have done so in piecemeal fashion as successively more sophisticated capabilities have been acquired and put into practice. The process of acquisition has clearly been one of purposive effort involving a succession of incremental steps, with production capabilities being developed somewhat in advance of investment capabilities. The selectivity of import substitution for the elements of technology has meant continued reliance on imports for at least some elements in almost all industries, but the pattern of imports has continually shifted as local capabilities have replaced foreign ones and as new industries have been developed. In turn, the selectivity of import substitution is complemented in the pattern of exports by specialization among the elements of technology in line with what one would expect to be South Korea's comparative advantage. And there is evidence that the pattern of exports, like the pattern of imports, has been shifting rapidly in response to the country's fast-paced acquisition of technological capability in many areas. What is perhaps most remarkable in this respect is that export activity appears to be an important vehicle for acquiring additional technological capabilities.

South Korea's ability to industrialize without extensive reliance on proprietary transfers of technology is in part explained by the nature of technology and product differentiation in the industries on which its growth has so far crucially depended. Many of these industries – such as plywood, textiles, and apparel – use relatively mature technologies; in such cases, mastery of well-established and conventional methods, embodied in equipment readily available from foreign suppliers, is sufficient to permit efficient production. The products of many of these industries are either quite highly standardized (plywood, for example) or differentiated in technologically minor respects and not greatly dependent on brand recognition for purchaser acceptance (textiles and apparel, for example). Thus, in most of the industries that have been intensively developed, few advantages are to be gained from licensing or direct foreign investment as far as technology acquisition and overseas marketing are concerned.

Nonetheless, exceptions exist, most notably in the chemical industry, where South Korea has had to rely extensively on direct foreign investment to establish and expand production, no doubt because of the reluctance of the technology suppliers to transfer technology via other modes. But in other industries where technology is also proprietary, a number of examples attest to the fact that South Korean industry has managed to initiate, and in most cases to operate successfully, a variety of high-technology industrial activities by means of licensing and turnkey arrangements. To cite two cases: arrangements of this kind were used to acquire the most modern shipbuilding technology in the world, and to incorporate the most recent technological advances in the integrated steel mill.

More generally, the country's recent experience in promoting technologically sophisticated industries indicates that their development may involve greater reliance on licensing as a way of acquiring technology. It is also probable that the shift to new industries may imply greater dependence on direct foreign investment. Indeed, greater dependence on such proprietary transfers of technology is observed starting in the latter half of the 1970s. But for the period in which semi-industrial status was achieved – that is, from the mid-1960s to the mid-1970s – there can be no doubt about the overwhelming role of indigenous effort in using capital goods imports and informal transfers of technology to acquire technological capability.

What is unique about the South Korean experience is not the importance of indigenous effort to assimilate technology. As we indicated at the outset, research on all semi-industrial economies shows that such effort is crucial for the acquisition of technological capability. What is unique about the South Korean experience is the speed and effectiveness of acquisition and the interplay between technological development and trade in the elements of technology. Near the beginning of the chapter, we outlined a framework for understanding and analyzing that interplay. The framework, like everything known about technological development, suggests the importance of paying attention to fine distinctions among technological capabilities when trying to understand the role of experience-based efforts in the cumulative buildup of technological capability. Because of space limitations, our discussion of South Korean technological development in later sections glossed over some of these distinctions.

Although we have enough details to conclude that such distinctions are fundamental, we lack a sufficient base of knowledge to generalize from South Korea's experience about many key aspects of formulating a strategy for technological development. We would emphasize four areas where more knowledge needs to be sought in order to understand the sources of South Korean success. One is the positive relationship between education and technological development – the details of this relationship remain to be uncovered. Another is the difference between competition in domestic and export markets in stimulating technological effort. We suspect but cannot yet demonstrate that an export-promoting policy regime provides a far more effective stimulus than either an inward-looking regime or an outward-looking one that simply aims at international competitiveness.

The third area is the role of government policy interventions in fostering South Korea's export-led industrialization and in stimulating technological development more generally. We know enough to conclude that the government's role cannot be characterized or judged in simple terms. We have not provided much discussion of the government's role because we wanted to focus on describing the main features of the country's technological development. Another reason is that more needs to be known before many conclusions can be reached. The last area for study is the importance of "initial conditions" in the early 1960s. Does rapid, sustained technological development in industry require the prior attainment of some minimal level of industrial competence built up through a process that is either much less rapidly paced or far more dependent on foreign production capability obtained through direct foreign investment or the use of expatriate manpower? The question remains wide open. Combining further study in these areas with the findings of complementary research on other semi-industrial economies may move us closer to answering important questions about the formulation of strategies for technological development.

NOTES

1. This section builds on Carl J. Dahlman and Larry E. Westphal, "Technological Effort in Industrial Development – An Interpretive Survey of Recent Research," in *The Economics of New Technology in Developing Countries*, ed. Frances Stewart and Jeffrey James (Boulder, Colo.: Westview Press, 1982), 105-37; and Dahlman and Westphal, "The Transfer of Technology: Issues in the Acquisition of Technological Capability by Developing Countries," *Finance & Development* 20 (December 1983), 6-9. The former paper uses *mastery* in much the same sense that *capability* is used here. Nelson's overview of issues in technological development also provides a pertinent perspective from which to understand the framework expounded in this section. Richard Nelson, "Research on Productivity Growth and Productivity Differences: Dead Ends and New Departures," *Journal of Economic Literature* 19 (September 1981), 1029-64.

2. For supporting argument and a survey of the relevant empirical evidence on these points, see Dahlman and Westphal, "Technological Effort in Industrial Development."

3. Adaptations have been observed to take place through changes that stretch the capacity of existing plants, break bottlenecks in particular processes, improve the use of by-products, adjust to new input sources, alter the product mix, and introduce a wide variety of incremental improvements in processes and product designs. For summaries of the most extensive program of case history research to date, see Jorge Katz, "Technological Change, Economic Development and Intra and Extra Regional Relations in Latin America," IDB/ECLA/UNDP Research Program on Scientific and Technological Development in Latin America, Working Paper no. 30 (Buenos Aires: United Nations Economic Commission for Latin America, 1978); Katz, "Domestic Technology Generation in LDCs: A Review of Research Findings," IDB/ECLA/UNDP Research Program on Scientific and Technological Development in Latin America, Working Paper no. 35 (Buenos Aires: United Nations Economic Commission for Latin America, 1980); Katz, "Technological Change in the Latin American Metalworking Industry: Results of a Program of Case Studies," IDB/ECLA/UNDP Research Program on Scientific and Technological Development in Latin America, Working Paper no. 51 (Buenos Aires: United Nations Economic Commission for Latin America, 1982); and Simon Teitel, "Creation of Technology within Latin America," *Annals of the American Academy of Political and Social Science*, 458 (November 1981), 136-50.

4. Rosenberg forcefully demonstrates this from the perspective of economic history. Nathan Rosenberg, *Perspectives on Technology* (Cambridge: Cambridge University Press, 1976).

5. Sang Chul Suh, *Growth and Structural Changes in the Korean Economy, 1910-1940* (Cambridge, Mass.: Harvard University Press for the Council on East Asian Studies, Harvard University, 1978), p. 153.

6. Edward S. Mason et al., *The Economic and Social Modernization of the Republic of Korea*, Studies in the Modernization of the Republic of Korea, 1945-1975, Harvard East Asian Monographs, vol. 92 (Cambridge, Mass.: Harvard University Press for the Council on East Asian Studies, Harvard University, 1980), p. 449.

7. For details, see George M. McCune, *Korea Today* (Cambridge, Mass.: Harvard University Press, 1950), chaps. 3 and 8.

8. For details about American aid, see Anne O. Krueger, *The Developmental Role of the Foreign Sector and Aid*, Studies in the Modernization of the Republic of Korea, 1945-75, Harvard East Asian Monographs, vol. 87 (Cambridge, Mass.: Harvard University Press for the Council on East Asian Studies, Harvard University, 1979) and Mason et al., *The Economic and Social Modernization of the Republic of Korea*.

9. For information about these reforms, see Larry E. Westphal, "The Republic of Korea's Experience with Export-Led Industrial Development," *World Development* 6 (March 1978), 347-82, and Mason et al., *The Economic and Social Modernization of the Republic of Korea*.

10. Much of this section is based on Larry E. Westphal, Yung W. Rhee, and Garry Pursell, "Korean Industrial Competence: Where It Came from," World Bank Staff Working Paper no. 469 (Washington, D.C.: World Bank, 1981), and Larry E. Westphal, Yung W. Rhee, Linsu Kim, and Alice Amsden, "Exports

of Capital Goods and Related Services from the Republic of Korea," World Bank Staff Working Paper no. 629 (Washington, D.C.: World Bank, 1984).

11. Westphal et al., "Korean Industrial Competence."

12. See Westphal et al., "Korean Industrial Competence," pp. 38 ff.

13. Kim provides a model of the underlying process of technological development and illustrates it with reference to South Korea's electronics industry. Linsu Kim, "Stages of Development of Industrial Technology in a Developing Country: A Model," *Research Policy* 9 (July 1980), 254-77.

14. Yung W. Rhee and Larry E. Westphal, "A Micro Econometric Investigation of Choice of Technology," *Journal of Development Economics* 4 (September 1977), 205-37.

15. On international subcontracting, see Westphal et al., "Korean Industrial Competence," pp. 52 ff.

16. Alice H. Amsden and Linsu Kim, "Korea's Technology Exports and Acquisition of Technological Capability" (Washington, D.C.: Productivity Division, Development Research Department, World Bank, 1982), second draft.

17. Additional technological histories – for petrochemicals, paper products, nylon yarn and cord, and the integrated manufacture of basic steel products – displaying the same pattern appear in Seoul National University, "The Absorption and Diffusion of Imported Technology in Korea" (Seoul: Institute of Economic Research, Seoul National University, 1980), draft, and in John Enos, "The Choice of Techniques vs. the Choice of Beneficiary: What the Third World Chooses," in *The Economics of New Technology in Developing Countries*, ed. Stewart and James, 69-81.

18. Linsu Kim, "Technological Innovations in Korea's Capital Goods Industry: A Micro Analysis," ILO Working Paper WEP 2-22/WP 92 (Geneva: International Labor Office, 1982).

19. Linsu Kim and Youngbae Kim, "Innovation in a Newly Industrializing Country: A Multiple Discriminant Analysis," *Management Science*, forthcoming.

20. Much of this section is taken without further citation from Westphal et al., "Exports of Capital Goods and Related Services."

21. See Bela Balassa, "Development Strategies and Economic Performance: A Comparative Analysis of Eleven Semi-Industrial Economies," in *Development Strategies in Semi-Industrial Economies*, Bela Balassa et al. (Baltimore: Johns Hopkins University Press for the World Bank, 1982), 38-62; Gershon Feder, "On Exports and Economic Growth," *Journal of Development Economics* 12 (February-April 1983), 59-74; and Mieko Nishimizu and Sherman Robinson, "Trade Policies and Productivity Change in Semi-Industrialized Countries," Development Research Department Discussion Paper no. 52 (Washington, D.C.: World Bank, 1983).

22. See Carl J. Dahlman and Francisco Sercovich, "Exports of Technology from Semi-Industrial Economies and Local Technological Development," *Journal of Development Economics* 16 (September-October 1984), 63-99.

23. Amsden and Kim, "Korea's Technology Exports."

24. See Jinjoo Lee, "Development of Engineering Consultancy and Design Capability in Korea," in *Consulting and Engineering Design in Developing*

Countries, ed. Alberto Araoz (Ottawa: International Development Research Center, 1981), 61-78.

25. See Alberto Araoz, "Consulting and Engineering Design Organizations in Developing Countries," in *Consulting and Engineering Design in Developing Countries*, ed. Alberto Araoz (Ottawa: International Development Research Center, 1981), 9-52; Martin Brown, "Engineering of Industrial Projects: Some Reflections on Current Development Center Research," Industry and Technology Occasional Paper no. 11 (Paris: OECD Development Center, 1976), document no. DC/TI(76)4; Martin Brown and Jacques Perrin, with the assistance of Dominique Genet, "Engineering and Industrial Projects: A Survey of Engineering Service Organizations" (Paris: OECD Development Center, 1977), document no. CD/R(77)2; and David J. Teece, *The Multinational Corporation and the Resource Cost of International Technology Transfer* (Cambridge, Mass.: Ballinger, 1976).

26. See Araoz, "Consulting and Engineering Design Organizations"; Brown, "Engineering of Industrial Projects"; Brown and Perrin, "Engineering and Industrial Projects"; Teece, *The Multinational Corporation*.

27. Amsden and Kim, "Korea's Technology Exports."

REFERENCES

Abegglen, James C., and William V. Rapp. "The Competitive Impact of Japanese Growth." In Jerome B. Cohen, ed., *Pacific Partnership: United States–Japan Trade*, pp. 19-50. Lexington, Mass.: Lexington Books for the Japan Society, 1972.

Amsden, Alice H., and Linsu Kim. "Korea's Technology Exports and Acquisition of Technological Capability." Washington, D.C.: Productivity Division, Development Research Department, World Bank, 1982. Second draft.

Araoz, Alberto. "Consulting and Engineering Design Organizations in Developing Countries." In Alberto Araoz, ed., *Consulting and Engineering Design in Developing Countries*, pp. 9-52. Ottawa: International Development Research Center, 1981.

Balassa, Bela. "Development Strategies and Economic Performance: A Comparative Analysis of Eleven Semi-Industrial Economies." In Bela Balassa and Julio Berlinski, Ow Chin Hock, Thomas L. Hutcheson, Kwang-Suk Kim, T. H. Lee, Kwo-shu Liang, Daniel M. Schydlowsky, Zvi Sussman, Augustine H. H. Tan, and Larry E. Westphal, *Development Strategies in Semi-Industrial Economies*, pp. 38-62. Baltimore: Johns Hopkins University Press for the World Bank, 1982.

Brown, Martin. "Engineering of Industrial Projects: Some Reflections on Current Development Center Research." Industry and Technology Occasional Paper no. 11, document no. DC/TI(76)4. Paris: OECD Development Center, 1976.

Brown, Martin, and Jacques Perrin, with the assistance of Dominique Genet. "Engineering and Industrial Projects: A Survey of Engineering Service Organizations." Document no. CD/R(77)2. Paris: OECD Development Center, 1977.

Dahlman, Carl J., and Francisco Sercovich. "Exports of Technology from Semi-Industrial Economies and Local Technological Development." *Journal of Development Economics* 16 (September-October 1984): pp. 63-99.

Dahlman, Carl J., and Larry E. Westphal. "Technological Effort in Industrial Development – An Interpretive Survey of Recent Research." In Frances Stewart and Jeffrey James, eds., *The Economics of New Technology in Developing Countries*, pp. 105-37. Boulder, Colo.: Westview Press, 1982.

____. "The Transfer of Technology: Issues in the Acquisition of Technological Capability by Developing Countries." *Finance & Development* 20 (December 1983): pp. 6-9.

Enos, John. "The Choice of Technique vs. the Choice of Beneficiary: What the Third World Chooses." In Frances Stewart and Jeffrey James, eds., *The Economics of New Technology in Developing Countries*, pp. 69-81. Boulder, Colo.: Westview Press, 1982.

Feder, Gershon. "On Exports and Economic Growth." *Journal of Development Economics* 12 (February/April 1983): pp. 59-74.

Hall, G. R., and R. E. Johnson. "Transfers of United States Aerospace Technology to Japan." In Raymond Vernon, ed., *The Technology Factor in International Trade*, pp. 305-58. New York: National Bureau of Economic Research, distributed by Columbia University Press, 1970.

Hayami, Yujiro, and Vernon W. Ruttan. *Agricultural Development: An International Perspective*. Baltimore: Johns Hopkins Press, 1971.

Katz, Jorge. "Domestic Technology Generation in LDCs: A Review of Research Findings," IDB/ECLA/UNDP Research Program on Scientific and Technological Development in Latin America, Working Paper no. 35. Buenos Aires: United Nations Economic Commission for Latin America, 1980.

____. "Technological Change, Economic Development and Intra and Extra Regional Relations in Latin America." IDB/ECLA/UNDP Research Program on Scientific and Technological Development in Latin America, Working Paper no. 30. Buenos Aires: United Nations Economic Commission for Latin America, 1978.

____. "Technological Change in the Latin American Metalworking Industry: Results of a Program of Case Studies." IDB/ECLA/UNDP Research Program on Scientific and Technological Development in Latin America, Working Paper no. 51. Buenos Aires: United Nations Economic Commission for Latin America, 1982.

Kim, Linsu. "Stages of Development of Industrial Technology in a Developing Country: A Model." *Research Policy* 9 (July 1980): pp. 254-77.

____. "Technological Innovations in Korea's Capital Goods Industry: A Micro Analysis." ILO Working Paper WEP 2-22/WP 92. Geneva: International Labor Office, 1982.

Kim, Linsu, and Youngbae Kim. "Innovation in a Newly Industrializing Country: A Multiple Discriminant Analysis." *Management Science*, forthcoming.

Krueger, Anne O. *The Developmental Role of the Foreign Sector and Aid*. Studies in the Modernization of the Republic of Korea, 1945-75, Harvard East Asian Monographs, vol. 87. Cambridge, Mass.: Harvard University Press for the Council on East Asian Studies, Harvard University, 1979.

Lee, Jinjoo. "Development of Engineering Consultancy and Design Capability in Korea." In Alberto Araoz, ed., *Consulting and Engineering Design in Developing Countries*, pp. 61-78. Ottawa: International Development Research Center, 1981.

McCune, George M. *Korea Today*. Cambridge, Mass.: Harvard University Press, 1950.

Mason, Edward S., Mahn Je Kim, Dwight H. Perkins, Kwang-Suk Kim, and David C. Cole. *The Economic and Social Modernization of the Republic of Korea*. Studies in the Modernization of the Republic of Korea, 1945-75, Harvard East Asian Monographs, vol. 92. Cambridge, Mass.: Harvard University Press for the Council on East Asian Studies, Harvard University, 1980.

Nelson, Richard. "Research on Productivity Growth and Productivity Differences: Dead Ends and New Departures." *Journal of Economic Literature* 19 (September 1981): pp. 1029-64.

Nishimizu, Mieko, and Sherman Robinson. "Trade Policies and Productivity Change in Semi-Industrialized Countries." Development Research Department Discussion Paper no. 52. Washington, D.C.: World Bank, 1983.

Rhee, Yung W., and Larry E. Westphal. "A Micro Econometric Investigation of Choice of Technology." *Journal of Development Economics* 4 (September 1977): pp. 205-37.

Rosenberg, Nathan. *Perspectives on Technology*. Cambridge: Cambridge University Press, 1976.

Salter, W. E. G. *Productivity and Technical Change*. Cambridge: Cambridge University Press, 1960.

Seoul National University. "The Absorption and Diffusion of Imported Technology in Korea." Seoul: Seoul National University, Institute of Economic Research, 1980. Draft.

Suh, Sang Chul. *Growth and Structural Change in the Korean Economy, 1910-1940*. Cambridge, Mass.: Harvard University Press for the Council on East Asian Studies, Harvard University, 1978.

Teece, David J. *The Multinational Corporation and the Resource Cost of International Technology Transfer*. Cambridge, Mass.: Ballinger, 1976.

Teitel, Simon. "Creation of Technology within Latin America." *Annals of the American Academy of Political and Social Science* 458 (November 1981): pp. 136-50.

Westphal, Larry E. "The Republic of Korea's Experience with Export-Led Industrial Development." *World Development* 6 (March 1978): pp. 347-82.

Westphal, Larry E., Yung W. Rhee, Linsu Kim, and Alice Amsden. "Exports of Capital Goods and Related Services from the Republic of Korea." World Bank Staff Working Paper no. 629. Washington, D.C.: World Bank, 1984.

Westphal, Larry E., Yung W. Rhee, and Garry Pursell. "Korean Industrial Competence: Where It Came From." World Bank Staff Working Paper no. 469. Washington, D.C.: World Bank, 1981.

Technology Analysis & Strategic Management, Vol. 1, No. 1, 1989 29

Innovation Strategy of Local Pharmaceutical Firms in Korea: A Multivariate Analysis

YOUNGBAE KIM, LINSU KIM & JINJOO LEE

ABSTRACT *This article examines 37 local pharmaceutical firms in Korea to identify different patterns of innovation behaviour associated with four types of firms, which are categorized by two variables: the scale of operation and technological capability. Both bivariate and multivariate analyses support the hypotheses that these four types of firms exhibit different innovation strategies and in turn the level of performance they achieved. Specifically, large firms with high technological capability diversify their efforts, investing in their own R & D and tapping foreign capabilities, and as a result exhibit the highest degree of innovativeness, while large firms with low technological capability resort primarily to the transfer of foreign technology for short-term profitability. In contrast, small firms with high technological capability rely mainly on their own R & D efforts with the assistance from local R & D institutes and enjoy the highest growth rate, whereas small firms with low technological capability imitate technologically low-grade products by hiring experienced technical personnel from other firms. There are also fragmented indications that many local firms have long evolved from small firms with low capability to large firms with high capability. Finally, managerial and theoretical implications for the innovation strategies in LDCs/NICs are discussed.*

Introduction

Managing technological innovation has become an imperative for the firms in newly industrializing countries (NICs), as they attempt to shift their focus on technologically sophisticated industries in the face of the recent erosion of their competitive advantage in labour-intensive industries. In the former where technological dynamics are a driving force of market competitiveness, firms in developed countries (DCs) show increasing reluctance to transfer technologies to NICs that have a potential to become competitors in the international markets. Thus, how to develop technology for new products and processes appears to be a crucial strategic decision for survival and growth of the firms in NICs in these arenas.

It has long been recognized in DCs as important to manage technological innovation from the strategic perspective: matching the firm's strategy to environ-

Youngbae Kim is Assistant Professor of Management Science, Korea Advanced Institute of Science & Technology, PO Box 150, Chongryang, Seoul 130–650, Korea (Tel: 967–8901 ext. 3861, Fax: 963–4013, Telex: K27380); Linsu Kim is Professor of Management, College of Business Administration, Korea University, Seoul, Korea; Jinjoo Lee is Professor of Management Science, Korea Advanced Institute of Science and Technology, Seoul, Korea.

mental characteristics and its internal capabilities.[1,2,3] While much has been studied about the firm's management of technology in DCs—such as in-house R & D, internal venturing, venture capital and nurturing, licensing, collaborative R & D, informal know-how trading, R & D subcontracting, acquisition, and merger[4,5,6,7,8]—little systematic research on innovation strategies has been done in less developed countries (LDCs) or in NICS with a few exceptions.[9,10,11]

This study, with the Korean pharmaceutical firms as the case in point, examines how they have managed to innovate and improve products and processes and have successfully competed against firms from DCs.

The pharmaceutical industry is highly research-oriented, particularly in long-term basic and applied research. Furthermore, the industry requires enormous expenditures in brand image-building and sales promotion. These industry characteristics lead multinational firms originated from a few DCs to dominate the world drug market. For example, multinational pharmaceutical firms accounted for 78% of the drug market in Brazil, 53% in Argentine, and 70% in India.[12] However, the local pharmaceutical firms in Korea, unlike in other LDCs/NICs where multinationals dominate the market, have managed to grow technologically and commercially to account for almost 90% of the Korean drug market, which ranks 15th in size in the world.[13]

This article first presents the modes of technology development in LDCs/NICs and then develops a research scheme to identify different patterns of technology development behaviour among different types of local firms[14] in the Korean pharmaceutical industry. Both bivariate and multivariate methods are used to analyse data from the sample of 37 local firms.

Technology Development Modes in LDCs/NICs

There are many ways whereby firms in LDCs/NICs can obtain technology for new product development and productivity improvement. Lee, Bae & Choi[15] classified the major technology development strategies available to local firms in LDCs/NICs into the following: (1) formal technology transfer (hereinafter, formal T/T); (2) non-formal technology transfer (hereinafter, non-formal T/T); and (3) indigenous R & D. These alternative strategies have both advantages and disadvantages in terms of implementation costs, likelihood of commercial success, time to fruition, and amount of expected market returns.[16]

First, given the weak technological base and the vast pool of foreign technological knowledge available, local firms resort heavily to technology transfer from DCs through various formal mechanisms. *Formal T/T* includes all the market-mediated mechanisms for international technology transfer with formal contracts and explicit payments such as patents and know-how licensing, the purchase of turnkey plants, and technical consultancies. While the formal T/T has many advantages,[17] it is often neither easy nor costless to search for suitable foreign technology sources, and costs associated with the use of proprietary know-how are not trivial. In general, payments for technology transfer are moderately high and such restrictions as tie-in clauses (a requirement that the licensee purchase raw materials and equipment from a source approved by the foreign licensor), export and sales restraints, and quality and design control are frequently required.

Second, local firms can have access to foreign technology without explicit payment for the transfer. *Non-formal T/T* includes purchasing of capital goods and components, consulting technical publications, and copying or reverse engineering of foreign products. Foreign buyers (especially in the case of original equipment manufacturer (OEM) exports) also provide valuable free technical assistance.[18] The primary advantage of non-formal T/T lies in the low cost of its implementation with accumulation of technological experience to some extent. However, this strategy is neither possible nor profitable, when the technology in question is quite complex and tacit, or well protected by patent law or trade secrecy.[19,20,21,22]

Finally, they may generate new products and processes through *indigenous R & D* efforts. In the process of indigenous R & D activity in LDCs/NICs, the role of local research institutes tends to be of particular importance. Among the roles of local research institutes are: (1) implementation of the firm-financed commissioned projects, in which a firm's own R & D personnel get involved to make the transfer of research results easier; (2) participation in joint advanced research with the local firm, financed jointly by the firm and government; and (3) provision of experienced R & D personnel to local firms. Indeed, research institutes in LDCs/NICs may be in a unique position to synthesize the needs of industrial firms and existing technology to bring forth technological innovation.[23]

Indigenous R & D efforts in LDCs/NICs have assimilated and/or improved imported foreign technologies, leading to the accumulation of technological capability. In the face of increasing technology protectionism prevailing in DCs, indigenous R & D becomes an important source of technology development in LDCs/NICs. Yet, indigenous R & D is not always the most effective way of initially developing all of the necessary elements of a technology, because it requires a very long time for fruition with significant financial and human resources and typically involves high risks of failure.[24]

These alternative strategies for technology development are not mutually exclusive. Rather, local firms use them in combination to accelerate the acquisition of technological capability. What is important is to understand the effectiveness of these strategies in different firms under different situations, since the appropriate strategy may depend on the technological and market characteristics of the particular industries under consideration, and on the availability of resources, managerial preferences, and corporate goals of the individual firms.

Research Scheme

The effectiveness of a firm's strategy with respect to technological innovation can be influenced by many factors, both internal and external. The external factors include market structure (e.g., degree of product differentiation, type of entry barriers and their height), the nature and technological characteristics of the industry (e.g., consumer goods vs. capital goods industry, product/market evolution stages, and operation technology type), and government policies (anti-trust policies and other regulations), just to name a few.[25,26,27,28,29,30] These environmental factors tend to differentiate the primary thrusts of technology development patterns among various sectors of industry.

Technology development strategy may also be expected to differ among firms within the same industrial sector due to substantial differences in their internal factors. The internal factors include the firm's scale of operation, financial capability, technological capability, and its bargaining power in acquiring foreign technology. In addition the internal factors also include its strategy with respect to market competition, the degree and nature of export orientation, the entrepreneurial orientation of the management and engineering leadership, the availability of engineering personnel, and its market share. Several studies in DCs[31] and in Korea[32] suggest that these factors are highly interrelated with each other. For example, compared to a small firm, the large firm is likely to have a large scale of operation with greater financial and human resources,[33] a larger market share therefore with a different market strategy,[34] and a greater bargaining power against technology suppliers by exploiting and maintaining other alternative sources.[35]

Of the factors presented above, external factors may not vary much among firms when a single industry is examined as in this study. Thus, two internal factors—the firm size and technological capability—appear to provide a useful research scheme to categorize Korean pharmaceutical firms and to identify differences in innovation behaviour among them. The *scale of operation* or *firm size* encompasses several variables. It includes, as pointed out earlier, financial resources, bargaining power in acquiring foreign technology, market share, and the size of human resources. A difference in these respects is expected to account for a significant dissimilarity in innovation behaviour among firms. Particularly, the large firm is likely to have resources and networks to develop more diverse contacts with foreign sources and to have stronger bargaining power against foreign technology suppliers than the small firm. In consequence, the larger the firm, the more likely it will resort to foreign technology.

Technological capability is another important variable accounting for a firm's technological behaviour.[36] Technological capability can be defined as the ability to make effective use of technological knowledge in efforts to assimilate, use, adapt, change, or create technology. It has three elements; production capability (ability to operate and maintain production systems), investment capability (ability to design and erect new production systems), and innovation capability (ability to generate new products and processes).[37] The greater technological capability the firm has, the more effectively it will assimilate transferred technology (the technology with highly tacit components, in particular), the more likely it will also rely on its own technological efforts and the more technologically innovative it will be to respond to changing market competition.

Correlation of the two variables leads to a research scheme of four types of firms as shown in Figure 1: (1) large scale with high capability; (2) large scale with low capability; (3) small scale with high capability; and (4) small scale with low capability. There may be some correlation between the scale of operation and technological capability, as the large firm is likely to have more financial resources than the small one to invest in human resources and R & D; nonetheless, previous works have revealed that there are both technologically static and dynamic firms among both large and small ones within the same industry.[38]

Hypotheses

Large firms with high technological capability (Type I) are able to diversify their technological efforts, both by investing in their own R & D activity and by tapping foreign sources. Kim, Lee & Lee[39] observed in the Korean computer industry that the 'industry dominators' (similar to Type I firms) invest heavily not only in in-house R & D but also in establishing close relations with external sources such as

Scale of Operation	Technological Capability: High	Technological Capability: Low
Large	Type I (n=11) NE=1041 AGE=44.6 TC=0.31 TMR=9.1% HIMR=1.9% RDR=1.9%	Type II (n=9) NE=541 AGE=25.1 TC=−0.52 TMR=4.8% HIMR=0.5% RDR=0.3%
Small	Type III (n=7) NE=131 AGE=17.4 TC=0.98 TMR=10.7% HIMR=1.6% RDR=5.1%	Type IV (n=10) NE=123 AGE=12.2 TC=−0.55 TMR=3.5% HIMR=0.5% RDR=0.5%

Figure 1. Research scheme and sample characteristics.

Notes:

NE: *Number of Employees*

AGE: *Age of the Firm*

TC: *Technological Capability* is a composite measure derived from the linear summation of standardized factor scores of TMR, HIMR, and RDR, which are the outputs of a principal component analysis.

TMR: *Technical Manpower Ratio* refers to the ratio of employees with degrees in science/engineering to NE.

HIMR: *High Level Technical Manpower Ratio* refers to the ratio of employees with postgraduate degree in science/engineering to NE.

RDR: *R & D Investment Ratio* to total net sales.

foreign licensors and local R & D institutes which can provide critical technology elements. These firms tend to combine external technological inputs with in-house technological efforts to expedite the acquisition of technological capability. Given their aggressive strategy to diversify the sources of technology and ample financial and human resources, their degree of innovativeness in terms of the number of new products and patents they produced would be highest among the aforementioned four types of firms.

The large firms with low capability (Type II) are likely to rely on foreign multinationals for the provision of new products and production technologies through licensing agreements and technical assistance contracts. Although they may become dependent on foreign sources for technology, they may nevertheless enjoy first-mover advantages and establish monopolistic positions within certain domestic market arenas through the use of a licensed product with an identifiable patent and brand name. That is, instead of promoting indigenous R & D efforts, they would compete based on market promotion efforts of foreign licensed products, and may in turn result in low technological innovations. Similar patterns of innovation behaviour were also found in 'sales agents' firms in the Korean computer industry.[40]

In contrast, the small firms with high capability (Type III), paralleled with 'entrepreneurial innovators',[41] are likely to rely less on licensing foreign technology but maintain a close relationship with local R & D institutes and universities to obtain technology for the development of new products and processes. Since their competitive advantages stem primarily from aggressive indigenous development efforts with the assistance from local R & D institutes, their degree of innovativeness would be higher than Type II and Type IV firms.

Finally, the small firms with low capability (Type IV), having neither resources to develop ties with foreign multinationals nor its own technological capabilities, are likely to depend on a non-formal T/T strategy for development of technologically low-grade products. Imitation or reverse engineering of existing foreign technology may be their primary way of technological efforts. Foreign vendors that supply components and parts and domestic firms that have experience in using that foreign technology would be important sources of technological information.[42] This type of small firms has been found in many industries ('imitators' in Amsden & Kim;[43] 'petty imitators' in Kim, Lee & Lee[44]).

Based on the foregoing paragraphs, it can be hypothesized about innovation strategies in terms of technology development modes and the degree of innovativeness among four types of firms, as Hypothesis 1 in Table 1.

Table 1. Innovation Strategy and Performance by Four Types of Firms.

Scales of Operation **Technological Capability**	**Large** **High** *Type I*	**Large** **Low** *Type II*	**Small** **High** *Type III*	**Small** **Low** *Type IV*
H1. *Innovation Strategy*				
a. Technology Development Modes	Formal T/T[a] Indigenous R & D	Formal T/T	Indigenous R & D	Non-formal T/T[b]
b. Degree of Innovativeness	High	Low	High	Low
H2, *Economic Performance*				
a. Growth Rate	Medium	Low	High	Medium
b. Profitability	Medium	High	Low	Low

Notes:
[a] Formal technology transfer.
[b] Non-formal technology transfer.

With different organizational capabilities and different innovation strategies, four types of firms are expected to show different level of performance in terms of profitability and growth rate. Specifically, large firms (Type I & Type II) tend to be more profitable than the small firms (Type III & IV) in the pharmaceutical industry, because of their advantages in scale economies and foreign-licensed products whose marketability, in general, has been proven elsewhere. In terms of growth rate, the firms with high capability (Type I & III) are more likely to outperform the firms with low capability (Type II & IV), because technological innovation, which requires relatively long gestation periods, may increase the probability of long-term survival or growth at the expense of short-term profitability. These relationships are also presented as Hypothesis 2 in Table 1.

Research Methods

Data for this study were collected from 37 local pharmaceutical firms selected from the population of 317 minus 16 multinationals. Since the small firms with low technical capability account for more than 85% of the population in the industry, a probability sampling plan would result in overrepresentation of Type IV firms. As an alternative to maintain the sample variability in terms of firm size and technological capability, a probability sampling plan was used to sample ten firms out of the population of small firms with low technological capability and 27 firms were randomly selected from the rest of the population. The distribution and characteristics of the sample in terms of the four types of firms and classificatory variables are shown in Figure 1.

Instruments to identify sources of technology were first delineated from examples of the literature[45,46] and structured interviews with industry experts, and then tested in a pilot study of five firms. The revised instruments were used for interviews with the CEOs and heads of technology management (R & D and/or production managers). In addition, objective data on the number of new products and patents, and economic performance indicators were also collected.

In this study, *the scale of operation* or *firm size* was measured in terms of the number of employees. Fixed assets or sales values may also be used for the same purpose. Yet, correlations between the number of employees and these two variables are very high in this study (correlation coefficients are 0.97 and 0.96, respectively).

Technological capability, as mentioned before, refers to the ability to make effective use of technological knowledge in efforts to assimilate, use, adapt, change, or create new technology. Dahlman and Westphal[47] suggest technological capability can be acquired through training, education, experience, and technological efforts. Thus, technological capability may be measured by such surrogates as the number of engineers and R & D investments. Using a principal component analysis, this study constructed a composite measure for technological capability which is a linear combination of standardized factor scores of technological manpower ratio to total employment (TMR), high level technological manpower ratio to total employment (HIMR), and R & D investment ratio to total sales (RDR).

Figure 1 reports the average age of each group as well as the mean number of employees (NE) and the mean score of technological capability (TC). The average number of employees of large firms (Type I & II) is 1041 and 541 respectively, while small firms (Type III & IV) have less than 150 employees on an average. Technological capability scores of Type I & III (high capability) are also significantly higher than Type II & IV firms (low capability).

Results

There appear to be three ways to develop new products for the local pharmaceutical firm in LDCs/NICs. First, the local firm may develop the new product with raw materials developed through *its own R & D efforts* (indigenous R & D strategy), in which case it needs to have or develop capabilities related not only to the production of chemical intermediates but also to the formulation of these raw materials to produce the finished drug. Under the process patent system, the innovation efforts of local firms mainly focus on the development of substitution processes of the patented drug. Second, the local firm may rely on multinationals for both the raw materials and the technical knowledge needed to formulate them. Through *licensing* or *technical assistance agreements* (formal T/T strategy) the local firm usually obtains from multinationals the necessary elements, including documentation for formulation and quality control and the active component to produce the final product. Patents and trademarks are generally involved. Third, the local firm may even *purchase raw materials in the open market* (usually from non-patented sources such as Italy) to produce the finished drug (non-formal T/T strategy), for which formulation technique is easy to acquire and master.[48]

The data presented in Table 2 show how the four types of local firms differ in generating new drugs. Type I and II (large-scale) firms produced 49.4% and 68.6%, respectively, of the total number of new drugs through 'formal T/T' and as much as 31.0% and 27.7%, respectively, through 'non-formal T/T'. In contrast, Type III (small-scale/high capability) firms produced the majority (61.3%) of the new products through their 'indigenous R & D', while Type IV (small-scale/low capability) firms produced 97.3% through 'non-formal T/T'.

Table 2 also presents data that reflect the outcome of the different strategy of the four types of local firms. It shows that compared to Types II and IV (low capability) firms, those with high technological capability (Types I and III) introduced a greater number of new products (3.89 and 4.14 vs. 6.47 and 5.73), innovations (0.13 and 0.03 vs. 1.38 and 0.69), and patents (none vs. 0.71 and 0.05) per year. The more the firm strives to become technologically independent by investing in acquiring technological capability, the more new products, innovative products, and patents it generates.

Table 2. Innovation Strategies by Four Types of Firms.[a]

Scales of Operation / **Technological Capability** / *Variables*	**Large** **High** *Type I* *(n=11)*	**Large** **Low** *Type II* *(n=9)*	**Small** **High** *Type III* *(n=7)*	**Small** **Low** *Type IV* *(n=10)*	**Two-way ANOVA** *firm size*	*tech capabil*	*interaction*
NEW PRODUCTS[b] developed by (%)							
Indigenous R & D	19.6	3.7	63.1	0.4	***	***	***
Formal T/T[c]	49.4	68.6	0.4	2.3	***	**	*
Non-formal T/T[d]	31.0	27.7	36.5	97.3	***	***	***
DEGREE OF INNOVATIVENESS							
Number of New Products[b]	6.47	3.89	5.73	4.14		**	
Number of Innovations[e]	1.38	0.13	0.69	0.03	***	***	***
Number of Patents Applied	1.50	0.04	0.24	0.00	***	***	***
Number of Patents Acquired	0.71	0.00	0.05	0.00	***	***	***

Notes:

[a] Six year average between 1979 and 1984.

[b] New products are defined as products introduced for the first time in the firm. They may not necessarily contain significant technological change.

[c] Formal technology transfer.

[d] Non-formal technology transfer.

[e] Innovations, as contrast to new products, are defined as significant technological changes in products and processes. Thus, product innovations may be a subset of new products.

*** significance level <0.01
** significance level <0.05
* significance level <0.1

In order to probe technological behaviour further in detail, this study examines the relative importance of external technology sources among four types of firms. Table 3 presents both subjective and objective data on the source of technological inputs for innovation. It shows that most variables are statistically significant to differentiate among four types of firms, but any subjective score below 2 may be ignored, as it indicates that managers of the sample firms interviewed considered the source to be of no more than 'slight importance'. To discuss just highlights, Type I (large-scale/high capability) firms viewed the role of local research institutes, the consultancy of individual experts from local universities and R & D institutes, foreign licensors and those foreign firms that provide various technical assistance, and foreign publications as important sources of technological inputs for their efforts to generate and improve products and processes. Such a view is substantiated by objective data on the quantity of foreign technical contracts (licensing and assistance) and the number of contracts with local research institutes. The former was greater in Type I firms than in any other groups and the latter was relatively greater for Type I firms than firms in Types II and IV.

Type II (large-scale/low capability) firms, by contrast, took into account only the role of foreign firms and publications as highly important for their product generation, which is also substantiated by the number of technical contracts (both assistance and licensing) with foreign firms.

The two groups of small firms also exhibited different behaviour in obtaining technological inputs. Type III (small-scale/high capability) firms viewed the role of local R & D institutes and foreign publications as very important for their innovation efforts. Though they only represented one-eighth of Type I firms in terms of the number of employees, they had a greater average number of contracts with local research institutes (3.29 vs. 2.83).

In contrast, Type IV (small-scale/low capability) firms perceived the role of competitors, the hiring of experienced technical personnel from other firms, the role of foreign suppliers who often provide informal technical assistance, and foreign publications as important sources of their technological inputs. The first two variables are highly related. These firms hired from competitors key engineering personnel, who tap useful technical information from former organizations through informal communication channels in imitating products that are being phased out by competitors. Interestingly, all types of firms take advantage of foreign publications.

In summary, the four types of local firms use different strategy and in turn exhibit different behaviour related to technological innovation. Specifically, Type I (large-scale/high capability) firms diversify their efforts, investing heavily in their own R & D efforts and in their joint research with local R & D institutes while at the same time tapping foreign technological capabilities. As a result, their degree of innovativeness appears to be highest among the four types of firms. Type II (large-scale/low capability) firms resort solely to technology licensing with little indigenous R & D investments, while Type III (small-scale/high capability) firms rely mostly on their own R & D efforts (see the size of R & D investment in Figure 1) with the assistance of local R & D institutes. Type IV (small-scale/low capability) firms imitate existing low-grade products by hiring experienced technical personnel from competitors. Thus, the aforementioned hypothesis 1 concerning different patterns of innovation behaviour in terms of modes of technology development and the degree of innovativeness is supported.

Table 4 shows a few economic performance indicators of the four types of firms. Returns on both sales and investments are higher in large firms (Type I &

Table 3. Sources of Technology by Four Types of Firms.[a]

Scales of Operation Technological Capability *Variables*	Large High *Type I* *(n=11)*	Large Low *Type II* *(n=9)*	Small High *Type III* *(n=7)*	Small Low *Type IV* *(n=10)*	Two-way ANOVA *firm size*	 *tech capabil*	 *interaction*
SUBJECTIVE DATA[b]							
Importance of Local Sources							
research institutes	3.36	0.89	3.86	1.00		***	
suppliers	0.45	1.00	0.29	1.90		**	
competitors	0.27	1.00	0.71	3.70	***	***	***
government agencies[c]	0.00	0.56	0.43	0.00			*
individual consultants	2.55	1.00	1.71	1.00		*	
technical manpower hired	1.18	1.00	0.00	2.60		*	**
Importance of Foreign Sources							
technical assistance	3.73	4.44	0.57	0.70	***		
training program	3.55	4.22	0.00	0.00	***		
suppliers	1.73	1.67	0.43	2.60			*
publications	4.82	4.33	4.57	3.10	**	***	
overseas observation	1.00	1.89	0.71	0.50	*		
OBJECTIVE DATA							
Number of foreign assistance contracts	10.03	8.89	0.00	0.23	***		
Number of technical licensing contracts	2.18	0.78	0.00	0.00	***	*	*
Number of contracts with local research institutes (75–79)	1.10	0.11	1.29	0.00		***	
Number of contracts with local research institutes (80–84)	1.73	0.55	2.00	0.00		***	

Notes:
[a] Figures in the table indicate the mean value of each group.
[b] Subjective data are measured using 6 point Likert Scale; 0, unimportant, through 5, very important.
[c] Government agencies include government-sponsored information centres and technical assistance institutes, and industry associations.

*** significance level <0.01
** significance level <0.05
* significance level <0.1

II) than in small firms (Type III & IV), as hypothesized. Production growth is highest in Type III (small-scale/high capability) firms. But, contrary to the Hypothesis 2-b, the growth rate of Type I firms is lower than other types of firms. Since the performance level of firms can be influenced by various environmental and managerial variables,[49,50] it may be difficult to explain the performance difference among four types of firms only with their different organizational capabilities and innovation strategies. However, the results can be interpreted by the fact that large firms can take advantage of foreign licensed products, whose marketability was already proven elsewhere and quality tends to be perceived superior and thus preferred, even at a premium price level, by domestic consumers. Dependence on foreign technology without indigenous technological capability, nevertheless, may increase short-term profitability at the expense of long-term adaptability under the growing trend of 'technology protectionism' in DCs.

A multiple discriminant analysis was undertaken to see whether interval variables used for the bivariate analyses above differentiate the technological behaviour of the four groups in a multivariate setting. One might argue that the sample size of 37 firms within four groups may not be enough to assume

Table 4. Economic Performance by Four Types of Firms.[a]

Scales of Operation	Large	Large	Small	Small	Two-way ANOVA		
Technological Capability	High	Low	High	Low			
	Type I	*Type II*	*Type III*	*Type IV*	*firm*	*tech*	
Variables	*(n=11)*	*(n=9)*	*(n=7)*	*(n=10)*	*size*	*capabil*	*interaction*
Production Growth Rate (%)	20.4	27.0	52.9	36.1	***		*
Return on Sales	8.8	11.0	5.0	4.4	***		
Return on Investment (%)	31.6	38.0	17.6	25.2	***		

Notes:
[a] Three year average between 1981 and 1983.

*** significance level <0.01
* significance level <0.1

multivariate normality in distribution, but this technique is not particularly sensitive to minor violations of the assumption.[51] Table 5 summarizes the results of stepwise discriminant analysis. Seven of the initial 11 subjective variables (see Table 3) were selected for the multivariate analysis when Wilk's lambda reached the lowest level.

The first two functions produced a very high degree of separation among the four groups as indicated by the final canonical correlation (0.9389 and 0.8502, respectively) and Wilk's lambda (0.0205 and 0.1727, respectively). It shows that the first function (which has a high positive coefficient in foreign assistance and foreign training programme) separates Types III and IV not only from each other but also from Types I and II, which are closely located. On the other hand, the second function (which has a high positive coefficient in the role of local R & D institutes but a negative one in that of domestic competitors) separates Types I and III not only from each other but also from Types II and IV. This indicates that technology development patterns of local pharmaceutical firms in this study can be classified primarily by two independent (i.e., not mutually exclusive) approaches: 'apprentices' and 'local-developers'.[53] While firms with the apprentice approach quickly acquired technological capability through heavy reliance on foreign technological sources, firms with the local-development approach promoted their technological capability through collaborative technology development efforts with local R & D institutes, rather than through mere imitation.

The plot of group centroids and individual cases by the first two functions, as presented in Figure 2, supports the general picture of the bivariate analyses in Table 3 and provides a visual presentation of standardized canonical discriminant function coefficients in Table 5. These two functions account for 94.32% of the total variance. Figure 2 shows that 91.89% of the firms in this study were correctly classified, which is significantly higher than the chance proportion of 25.6%.

Summary and Discussion

This study attempted to identify the pattern of technological behaviour of local firms in the Korean pharmaceutical industry. The results of this study show that the four types of local firms categorized by the scale of operation and technological capability adopted different strategies with respect to acquiring technology

Table 5. Results of Discriminant Analysis (stepwise).

Canonical Discriminant Functions

Function	Eigenvalue	% of Variance	Cumulative %	Canonical Correlation
1	7.44	69.84	69.84	0.9389
2	2.61	24.48	94.32	0.8502
3	0.60	5.68	100.00	0.6139

After Function	Wilk's Lambda	Chi-squared	Degree of Freedom	Significance
0	0.0205	118.62	21	0.0000
1	0.1727	53.57	12	0.0000
2	0.6231	14.43	5	0.0131

Standardized Discriminant Function Coefficient (n=37)

Variables	Function 1	Function 2	Function 3
Local research institutes	−0.09347	0.68885	−0.77968
Local competitors	0.48546	−0.90993	−0.27931
Government agencies	−0.11857	0.53973	0.81199
Local technical manpower hired	0.09242	0.33345	−0.66913
Foreign technical assistance	−0.66971	−0.01611	−0.16173
Foreign training program	−0.65609	−0.50517	−0.11954
Foreign suppliers	0.15511	−0.35713	−0.37164

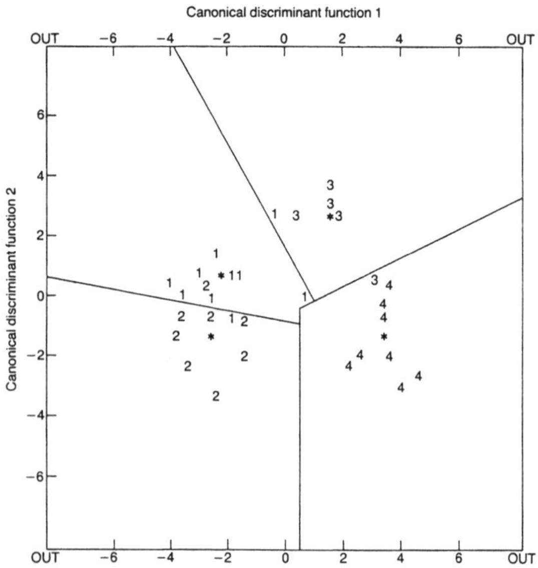

Figure 2. Two-function plot of centroids and cases.
Note: All-groups scatterplot—* indicates a group centroid.

and developing new products and processes, and in turn exhibited different levels of performance. These findings offer several implications.

First, there appear to be three primary ways, whereby local firms in LDCs/NICs can develop technology for new products and processes: formal T/T, non-formal T/T, and indigenous R & D. Certainly these alternative ways of technology development should not be viewed as mutually substitutable, rather complementary. Kim[53] asserts that it is important for firms in LDC/NICs to combine foreign technology transfer as a catalytic source for making a quantum jump in technological capability and indigenous R & D efforts as a source for absorbing and improving transferred foreign technology. That is, in order to improve expeditiously technological capability to strengthen their market competitiveness in technologically sophisticated industries, local firms in LDCs/NICs should search aggressively for every possible method of technology development.

Second, albeit important in combining indigenous R & D efforts and external technology sources, the innovation strategy of local firms in LDCs/NICs should be contingent, at least in the short term, upon their internal characteristics. That is, the management of local firms should take into account their organizational capabilities in choosing appropriate innovation strategies. For example, large firms with high technological capability should invest in their indigenous R & D activity, and at the same time exert efforts to develop effective ties with both foreign multinationals, which are relatively willing to transfer technology at a reasonable price, and local research institutes, which provide not only technological information but also trained R & D personnel. Relying on any one alternative technological source may lead to the deterioration of either short-term profitability or long-term adaptability. The small firms with low technological capability should focus their market niches in technologically low-grade product segments and actively search for available technology sources to which they have access without explicit payment of the transfer. The foreign products or diffusion of technology from customers, suppliers, and competitors may be their primary sources of technology for imitation or modification of existing products and processes. However, without an effort to improve significantly their technological capability these firms ultimately cannot grow to become a viable one in the competitive market.

Third, data discussed above merely present a cross-sectional picture of the technological behaviour of four types of local pharmaceutical firms in Korea. But data on the average age of the four types (44.6, 25.1, 17.4, and 12.2, respectively), as presented in Figure 1, lead one to raise questions regarding interrelations among them. Did Type I firms start from the beginning as Type I or have they evolved through other types? Will Type II firms grow to become Type I? What will happen to Type III and IV firms? Such questions can only be answered heuristically with in-depth historical case studies and by examining the future strategic direction of each firm under study. Data for the two-thirds of sample firms, for which in-depth case information is available, show that with a few exceptions about one-half of Type I forms have evolved from Type IV via Type II, while several others have evolved from Type IV via Type III. About a half of Type II and III firms have also grown from Type IV. According to interviews with sample firms, many Type II firms expressed their aspiration to build up their own technological capability like Type I firms, while Type III and IV firms also showed a similar aspiration to evolve into Type I firms. In short, there are many indications that the four types appear to show the growth path of local pharmaceutical firms in Korea.

This growth path of local firms may imply that the importance of various external information sources will change, as the firm evolves from one type to another. However, the recent environmental changes faced by Korean pharmaceutical firms make this projection difficult. The government's recent revision of the patent system and import liberalization will result in a significant environmental change for local pharmaceutical firms in Korea. Under such a new environment viable growth strategies of local firms will be significantly different from those reported in this article.

Fourth, a pattern emerged here appears to have both commonalities and differences from those found in previous studies in Korea. Kim, Lee & Lee[54] reported that firms in the Korean computer industry showed five different patterns in acquiring technological capability according to their innovation and production capabilities. Among them, industry dominators, sales agents, entrepreneurial innovators, and petty imitators, as mentioned before, are largely in parallel with Type I, II, III, and IV firms respectively in this study. This study, however, found no firms comparable to OEM producers in the computer industry (firms with high production capability but low innovation capability), which benefitted from the technical assistance from foreign OEM buyers. This stems from the fact that the pharmaceutical industry is not export-oriented and that production capability for cost efficiency for competitiveness in the international market is not as important as it is in the electronics and computer industries.

Kim and Lee[55] summarized the results of research undertaken in various industries in Korea to identify patterns of technological change and postulated that while dissimilarities originated from differences in firm size and operations technology deployed by different industries, similarities stemmed from the obsession of local firms for optimizing learning given resources. The results of this study largely confirmed this postulation. Yet, this study showed that the technological capability of the firm and sources of market competitiveness also appear to be important in accounting for differences in their technological behaviour among different firms in the pharmaceutical industry. This implies that local firms in LDCs/NICs should take into account the market and technological factors of the industries as well as their internal characteristics in choosing appropriate innovation strategies.

Finally, it is important to note that the pattern observed in the Korean pharmaceutical industry may not necessarily prevail in other newly industrializing countries where liberal policies toward direct foreign investments by multinational firms have been adopted, or in other developing countries which lack managerial and technical human resources. It is necessary, therefore, to replicate this study in different industries in different countries and incorporate the environmental variables in order to develop a more comprehensive theory of innovation strategies of local firms in LDCs/NICs.

Notes and References

1. B. Gold, 'Alternate strategies for advancing a company's technology', *Research Management*, *18*, 4, 1975, pp. 24–29.
2. R. L. Schill, 'Managing technology for competitive advantage in interindustry markets', *R&D Management*, *16*, 2, 1986, pp. 103–116.
3. B. Twiss, *Managing Technological Innovation*, 2nd ed., (London, Longman, 1980).

4. C. Freeman, *The Economics of Industrial Innovation*, 2nd ed., (Cambridge, MA, MIT Press, 1982).
5. J. Lowe & M. Silver, 'R & D strategies and variable demand', *R&D Management, 16*, 4, 1986, pp. 325–333.
6. W. Marcy, 'Acquiring technology—licensing do's and don'ts', *Research Management, 22*, 1979, pp. 18–21.
7. E. B. Roberts & C. A. Berry, 'Entering new businesses: selecting strategies for success', *Sloan Management Review, 26*, 3, 1985, pp. 3–17.
8. E. A. von Hippel, 'Cooperation between rivals: informal knowhow trading', Sloan School of Management Working Paper No. 1759-86 (Cambridge, MA, MIT, 1986).
9. Z. Bae & J. Lee, 'Technology development patterns of small and medium sized companies in the Korean machinery industry', *Technovation, 4*, 1986, pp. 279–296.
10. L. Kim & Y. Kim, 'Innovation in a newly industrializing country: a multiple discriminant analysis', *Management Science, 31*, 3, 1985, pp. 312–322.
11. L. Kim, J. Lee & J. Lee, 'Korea's entry into the computer industry and its acquisition of technological capability', *Technovation, 6*, 1987, pp. 277–293.
12. UN (United Nations), *Transnational Corporations in the Pharmaceutical Industry of Developing Countries*, ST/CTC/49 (New York, UN, 1984).
13. *Ibid.*
14. Local firms mean the firms without any foreign ownership. Foreign subsidiaries and joint ventures, thus, are excluded in the analysis of this study.
15. J. Lee, Z. Bae & D. Choi, 'Technological development process in a developing country with global perspective: a model', *R&D Management, 18*, 3, 1988, pp. 235–250.
16. Gold, *op. cit.*, Reference 1.
17. Marcy, *op. cit.*, Reference 6.
18. Bae & Lee, *op. cit.*, Reference 9; L. W. Westphal, Y. W. Rhee & G. Pursell, 'Foreign influences on Korean industrial development', *Oxford Bulletin of Economics and Statistics, 41*, 1979, pp. 359–388.
19. D. J. Teece, 'The market for know-how and the efficient international transfer of technology', *Annals of AAPSS, 458*, 1981, pp. 81–96.
20. D. J. Teece, 'Profiting from Technological Innovation: Implications for Integration, Collaboration, Licensing, and Public Policy', in: D. J. Teece (ed.), *The Competitive Challenge: Strategies for Industrial Innovation and Renewal* (Cambridge, MA, Ballinger, 1987).
21. E. A. von Hippel, 'Appropriability of innovation benefit as a predictor of the source of innovation', *Research Policy, 11*, 1982, pp. 95–115.
22. S. G. Winter, 'Knowledge and Competence as Strategic Assets', in: D. J. Teece (ed.), *The Competitive Challenge: Strategies for Industrial Innovation and Renewal* (Cambridge, MA, Ballinger, 1987).
23. J. M. Utterback, 'The role of applied research institutes in the transfer of technology to Latin America, *World Development, 3*, 9, 1975, pp. 665–673.
24. C. J. Dahlman & L. E. Westphal, 'The meaning of technological mastery in relation to transfer of technology', *Annals of AAPSS*, 458, 1981, pp. 12–26.
25. Gold, *op. cit.*, Reference 1; Twiss, *op. cit.*, Reference 3.
26. M. Porter, 'The Technical Dimensions of Competitive Strategy', in: R. S. Rosenbloom (ed.) *Research on Technological Innovation, Management, and Policy* (Greenwich, CT, JAI Press, 1983), pp. 1–33.
27. A. K. Chakravarti & W. E. Souder, 'Critical factors in technological innovation and their policy implications', *Technovation, 2*, 4, 1984, pp. 255–275.
28. M. I. Kamien & N. L. Schwartz, *Market Structure and Innovation* (New York, Cambridge University Press, 1982).
29. J. E. Ettlie, 'Organizational policy and innovation among suppliers to the food processing sector', *Academy of Management Journal, 26*, 1, 1983, pp. 27–44.

30. J. M. Utterback & W. J. Abernathy, 'A dynamic model of product and process innovation', *Omega*, *3*, 6, 1975, pp. 639–656.
31. Kamien & Schwartz, *op. cit.*, Reference 28; Ettlie, *op. cit.*, Reference 29.
32. A. H. Amsden & L. Kim, *Technology and Industrialization in Korea* (Development Research Department, the World Bank, 1985, a mimeograph).
33. R. Rothwell & W. Zegfeld, *Industrial Innovation and Public Policy: Preparing the 1980s and the 1990s* (Westport, CT, 1981, Greenwood Press).
34. C. Y. Woo & A. C. Cooper, 'Strategies of effective low share businesses', *Strategic Management Journal*, *2*, 3, 1981, pp. 301–318.
35. Amsden & Kim, *op. cit.*, Reference 32.
36. M. Fransman, 'Promoting technological capability in the capital goods sector: the case of Singapore', *Research Policy*, *13*, 1984, pp. 33–54.
37. C. J. Dahlman & L. E. Westphal, *Notes on the Conceptualization of Technology and Technological Capability* (Development Research Department, the World Bank, September, 1983, a mimeograph).
38. Kamien & Schwartz, *op. cit.*, Reference 28; Amsden & Kim, *op. cit.*, Reference 32.
39. Kim, Lee & Lee, *op. cit.*, Reference 11.
40. *Ibid.*
41. *Ibid.*
42. Kim & Kim, *op. cit.*, Reference 10.
43. Amsden & Kim, *op. cit.*, Reference 32.
44. Kim, Lee & Lee, *op. cit.*, Reference 11.
45. Kim & Kim, *op. cit.*, Reference 10; Westphal *et al.*, *op. cit.*, Reference 18.
46. T. J. Allen, D. B. Hyman & D. L. Pinkney, 'Transferring technology to the small manufacturing firm: a study of technology transfer in three countries', *Research Policy*, *12*, 1983, pp. 199–211.
47. Dahlman & Westphal, *op. cit.*, Reference 37.
48. D. Chudnovsky, 'The challenge by domestic enterprises to the transnational corporations' domination: a case study of the Argentine pharmaceutical industry', *World Development*, *7*, 1, 1979, pp. 45–58, also found a similar pattern in Argentina.
49. R. E. White & R. G. Hamermesh, 'Toward a model of business unit performance: an integrative approach', *Academy of Management Review*, *6*, 2, 1981, pp. 212–223.
50. K. J. Hatten, D. E. Schendel & A. C. Cooper, 'A strategic model of the U.S. brewing industry: 1951–1971', *Academy of Management Journal*, *21*, 4, 1978, pp. 592–610.
51. W. R. Klecka, *Discriminant Analysis* (New York, Sage University Press, 1980).
52. Kim first coined the 'apprentices' and 'imitators' approach to describe two distinct patterns in the acquisition of technological capability by Korean firms. 'Local-developers' are not identical to the 'imitators'. Both 'local-developers' and 'imitators' are technologically dynamic, but 'imitators' upgrade their technological capability through reverse engineering of existing foreign products locally available without formal assistance from external agencies, while 'local developers' not only exert their technological efforts but also resort heavily to local R & D institutes. L. Kim, 'Imitation and apprentice: the acquisition of technological capability in Korea' (College of Business Administration, Korea University, Seoul, 1987, a mimeograph).
53. L. Kim, 'The pros and cons of international technology transfer: a developing country view', paper presented at the second Annul IBEAR Research Conference, University of Southern California, 1988.
54. Kim, Lee & Lee, *op. cit.*, Reference 11.
55. L. Kim & H. Lee, 'Patterns of technological change in a rapidly developing country: a synthesis', *Technovation*, *6*, 1987, pp. 261–276.

[5]

Technovation, 6 (1987) 261-276
Elsevier Science Publishers B.V., Amsterdam — Printed in The Netherlands

Patterns of Technological Change in a Rapidly Developing Country: A Synthesis

Linsu Kim
College of Business Administration, Korea University, Seoul (Korea)

and Hosun Lee
Department of Management Science, Korea Advanced Institute of Science and Technology, Seoul (Korea)

ABSTRACT

Management of technological change has become increasingly important in many developing countries, as they evolve from agricultural to industrial societies. This paper attempts to integrate a series of recent studies undertaken in a rapidly developing country in order to identify patterns of technological change in manufacturing firms. The degree of complexity in operations technology and the scale of operations account for differences in their technological behavior among different firms in different industries.

1. INTRODUCTION

Technological change is known to have been a major determinant of national development. In developed countries (DCs), many studies have shown that more than 50% of long term economic growth stems from technological changes which improve productivity or lead to new products, processes, or industries [1]. For this reason, it is often questioned how science and technology can be effectively used for economic and social development in less developed countries (LDCs).

Many LDCs have made significant progress in industrial and technological development, evolving from dominantly agricultural, traditional societies to dominantly industrial, modern societies. In the process, the management of technological change has become increasingly important. The proper management of technological change at the enterprise and national levels requires the understanding in the pattern of technological change in the firm and industry[1].

0166-4972/87/$03.50

An early study on the electronics industry in Korea postulates a three-stage model that depicts the pattern of technological change in an LDC [2]. Such a model, based on one industry cannot, however, be generalized to other industries that have different characteristics. For this reason, Kim [2, p.273] concludes:

Many variables both at the industry and national levels may account for variations in the development pattern of industrial technology in different countries. The characteristics of production technology and market structure may be sources of variations at the industry level. Policy on export promotion and foreign investment, the availability of local entrepreneurs, the capability of the local scientific and engineering personnel, and the availability of natural resources may be important sources of variation at the national level.

Of these factors, production technology appears to be key for variations among industries in a given country. Woodward [3] in her seminal work on the impact of production technology upon organizational characteristics suggests that the degree of complexity in production technology[2] — small batch and unit production (hereinafter small batch), large batch and mass production (hereinafter large batch), and process production — dictates to a certain extent the behavior of the firm in undertaking technological innovation. In addition, Amsden and Kim [4] suggest that the scale of operation also accounts for variation in the behavior of technological change at the firm level, as the scale encompasses to a large extent financial and technological capability, the bargaining power against foreign technology suppliers, market share, and human resource capability, all of which have a strong bearing on firm's strategy toward its own R&D efforts and the acquisition of foreign technology. The fact that, in general, large firms produce sophisticated products whereas small firms produce unsophisticated products may also account for differences in technological behavior between the two groups of firms.

This paper is aimed at integrating recent studies undertaken by Kim and his associates in the industries that use all three types of operations technology — small batch technology [5,6], large batch technology [2,7], and process production technology [4,8], — to determine how and why the pattern of technological change varies by different production technology. The small batch operations usually produce highly differentiated products, whereas the process operations produce least differentiated products with the large batch operations in the middle. Such relations may be dipicted as in Table 1. Differences in technological behavior between large and small firms in the same production technology are also studied. The examination of the effect of both operations technology and the scale of operation on the pattern of technological change may lead to the development of a more general model than the one presented earlier. The subsequent sections will first describe the patterns of technological behavior exhibited by firms with different operations technologies. Then, an attempt will be made to synthesize these patterns and to explain why firms

TABLE 1

Technology complexity and product differentiation

		Degree of product differentiation		
		Low	Middle	High
Degree of technology complexity	Small batch and unit production			Machinery, Shipbuilding
	Large batch and mass production		Electronics, Automobile	
	Process production	Chemical, Pharmaceutical Cement, etc		

with different operations technologies and size show similar or dissimilar behaviors.

2. TECHNOLOGICAL CHANGE IN SMALL BATCH OPERATIONS

There are at least three studies on the behavior of technological change in industries with small batch operations. Amsden and Kim [5] undertook a case study of the three largest heavy machinery firms in Korea. Another study by Kim [9] examined, among other sectors, the shipbuilding industry which also deploys small batch operations. Kim and Kim [6] undertook an in-depth study of relatively small machinery firms that developed innovative products. These studies provide useful data to delineate the pattern of technological change between large and small firms in the industry with small batch operations.

Large machinery firms and giant shipbuilders relied heavily on foreign sources in the form of foreign licensing and technical consultancies for the initial installation of production processes and for product development. The machinery industry in the aggregate (ISIC 38) has accounted for a disproportionately large share of Korea's total foreign licenses. Between 1962 and 1981, it accounted for almost half of all technology imports (by number) but barely 10% of total domestic value added, even toward the end of the period. Of the total number of technologies licensed by the machinery sector, the general machinery subsector purchased more than half, despite the fact that it accounted for less than 10% of the value added of total machinery production [5]. Further, licenses in the general machinery are accounted for mostly by large firms [10]. The case study of large machinery firms and giant shipbuilders indicates that each of them has as many as 30–50 licences [4]. In recent years, these firms, however, have exerted increasingly their own R&D efforts to strengthen their own capability and in turn to reduce their reliance on foreign licensors, as they accumulate experience in production and product design.

In contrast to large firms that relied heavily on foreign licensings at the outset and then exerted their own efforts to assimilate imported technology in a short span of time, small firms in the industry with unit or small batch operation resorted primarily to their own limitative efforts to evolve organically over a long period of time. The case study undertaken by Amsden and Kim [4] reveals that small firms established their initial production facilities with primitive technologies developed by themselves and then gradually upgraded their production system and product quality through the imitative reverse engineering of foreign products and processes.

A more in-depth investigation undertaken by Kim and Kim [8], as presented in Table 2, reveals that four types of small, innovative machinery producers classified by two variables — the source of initiation for innovation (user or innovator) and the local availability of related foreign products (locally available or not) — exhibit distinctively different technological behavior. Table 2 shows, among other things, two most contrasting cases: the first and last columns in the table. When the innovation was triggered by either users' orders or serious inquiries and where related foreign products were locally available from which firms could learn and imitate creatively, firms used and benefited most from close interactions with users during both the idea generation and problem solving stages and from visits to the domestic users of such foreign products plus catalogs readily available at the local office of foreign suppliers. Consequently, such innovations were most imitative and least patented. In contrast, when the innovation was initiated by innovators searching for markets and where no related foreign products were locally available from which firms could learn, they used and benefited most from overseas observation of foreign suppliers and assistance of local R&D institutes. They also required more highly qualified technical personnel within the firm[3]. As a result, such innovations were least imitative and most patented. A multiple discriminant analysis of the same data also support the patterns which emerged in the bivariate analyses presented in Table 2 (see [6] for detailed results). Some of these small firms, however, made a new move in acquiring foreign technology. While they continue to imitate or adapt foreign products on the basis of observation and technical information obtained through informal channels, they have recently initiated technical licensings with foreign firms as a way to penetrate into the more sophisticated local and overseas markets [4].

To recapitulate, large firms with small batch operations relied heavily on the formal transfer of foreign technology primarily in the form of foreign licensings and consultancies for the initial erection of production system and for the design

TABLE 2

Difference of technological behavior by four types of small firms with unit batch operation

Factors affecting innovation	Related foreign products locally available		No related foreign products locally available	
	User-initiated Type I $n=21$	Innovator-initiated Type II $n=8$	User-initiated Type III $n=6$	Innovator-initiated Type IV $n=6$
*Results of cross tabulations (Percentage)**				
Entrepreneurs as:				
Key person in idea generation	59.1	71.4	66.7	66.7
Key person in problem solving	36.4	28.6	50.0	83.3
Firms used overseas observation	19.1	25.0	16.7	33.3
Firms obtained patents for NDIMs	5.0	12.5	16.7	66.7
Results of Kruskal-Wallis ANOVA (Mean Ranks)				
Educational level of key technical person	17.56	15.00	18.67	26.60
Position of key technical person within the firm	15.97	14.75	19.92	23.10
Customer's role in:				
idea generation+	24.64	15.00	12.33	11.90
problem solving+	26.02	18.05	14.83	7.30
Opportunity opened by import substitution policy	23.52	21.00	15.17	14.67
Contribution of domestic observation***	24.48	22.57	13.42	11.25
Firms obtained assistance from local R & D lab+	18.95	20.50	18.00	31.83
Usefulness of catalogs**	22.63	20.29	18.42	9.75
Degree of imitation+	27.42	16.25	13.92	9.67

*The figures reported in the first block is the proportion of firms that responded "yes". For example, 59.1% (first figure in the table) of the firms (Type I), whose innovation was triggered by users and who had related foreign products locally available, mentioned that it was the entrepreneur that played a key role in the initiation of innovation.

**significance level <0.1.

***significance level <0.05.

+significance level <0.01.

Source: Kim and Kim [6].

of their products, and then rapidly acquired technological capabilities through assimilating imported technology and through their own R&D efforts. Whereas, small firms using the same type of production technology developed their own primitive production systems and then grew slowly over a long period of time by imitative reverse engineering of foreign products. Informal mechanisms, such as personal observations, were a predominant form of foreign technology transfer for small firms.

3. TECHNOLOGICAL CHANGE IN LARGE BATCH OPERATIONS

Several studies examined the pattern of technological change in the industries that deploy large batch operations. Kim [2] and Kim and Utterback [7] postulate the evolution of technological change in the electronics industry, while Amsden and Kim [11] examined two major automobile producers in Korea[4].

Large firms in both the consumer electronics and automobile industries were also dependent on foreign technology but not as much as those with small batch operations. In the advent of an import substitution policy, large local firms which had previous trade contacts with foreign firms, signed technical assistance aggreements with leading foreign firms to import "packaged" technology so as to establish the initial production system. Such a technology transfer included assembly processes, product specifications, production know-how, training of local technical personnel at the technology supplier's site, service of foreign expatriates and even component parts. The immediate technological task was the implementation of transferred foreign technology for the assembly of imported components and parts, whose final products had been tested and proven elsewhere. For this purpose, only engineering (E) efforts were required.

Once the implementation task was accomplished, production and product design technologies were quickly diffused within the country, mainly through the mobility of local entrepreneurial and technical personnel from existing firms to new entrants. Competition stemming from new entrants, together with increasing export promotion, called for indigenous efforts to assimilate foreign technology, thus placing emphasis not only on engineering but also on limited development (D&E). Capabilities acquired through assimilation of foreign imported technology also enabled local firms to develoop, through imitative engineering, related products without foreign assistance. Then, the relatively successful assimilation of imported technology with the increased capability of local scientific and engineering personnel, led to the gradual improvement of foreign technology and the ability to implement their own novel ideas in limited areas through local indigenous efforts in research, development and engineering (R,D,&E). Table 3 summarizes the pattern of devel-

TABLE 3

The evolution of technological change in the industry with large batch and mass production technology

	Stage 1	Stage 2	Stage 3
New firms established through	Transfer of foreign technology	Mobility of local technical and entrepreneurial personnel	
Technical task emphasis	*Implementation* of imported technology	*Assimilation* for product diversification	*Improvement* for enhancing competitiveness
Critical human resources	Foreign experts	Local technical personnel trained at supplier firms	Local scientific and engineering personnel
Production technology	Inefficient		Relatively efficient
Predominant source of technological change	Transfer of 'packaged' foreign technology		Indigenous efforts
Predominant form of international of technology	'Packaged'		'Unpackaged'
Predominant sources of external influence on technological change	Supplier and government		Customer, competitor
Market	Local (low competition)		Local and overseas (high competition)
Emphasis on research, development, engineering	Engineering (E)	Development and Engineering (D&E)	Research, Development, and Engineering, (RD&E)
Supply source of components and parts	Mostly foreign		Mostly local
Important government policies	Import substitution and control of foreign investment		Export promotion
Role of local applied R&D	Consultative	Adaptive development	Research and development

Source: Kim [2].

opment of industrial technology in the electronics industry in Korea and critical conditions which stand out at different stages.

Small firms in the electronics and automobile industries deploy small batch technology rather than large batch system. However, small automobile producers, specializing in special-purpose trucks, resorted heavily to foreign licensings in order to get technical information and assistance for the production of sophisticated products, showing basically identical behavior with large automobile producers. In contrast, small electronics firms entered the industry by poaching experienced engineers from existing firms and cleaved to a market niche. Such a behavior is closer to that of small pharmaceutical firms to be elaborated subsequently under the subject of process production. In other words, there is no consistent behavior among small firms in these industries. The difference appears to stem largely from the gap between the degree of technological sophistication in products and the level of capability these firms had in tackling technological problems in producing such products. When the gap was large, the firm became dependent on foreign sources; otherwise, it became self-reliant.

To recapitulate, large firms with large batch technology, such as in the electronics and automobile industries, were initially foreign dependent to establish production processes and to produce products; but not as much as those with small batch technology, as the former produced a lesser number of more standardized products than the latter. These firms, however, exerted local efforts, as those with unit or small batch technology, to assimilate imported foreign technology by reversing the usual sequence of research, development and engineering. Small firms in the same industries, in contrast, deploy small batch technology to suit the quantity requirements. Some small firms that produce sophisticated products were initially foreign dependent, as were large ones, while other small firms have largely been self-reliant, like small machinery firms mentioned previously.

4. TECHNOLOGICAL CHANGE IN PROCESS PRODUCTION OPERATIONS

Two studies examined the pattern of technological change in the industries that use the process production technology. Amsden and Kim [4] undertook several case studies in the large cement, chemical, and steel firms and two small papermaking firms in Korea; while Kim, Kim, and Lee [8] examined the difference in technological behavior between large and small pharmaceutical firms in Korea.

Under the import substitution drive by the government, the initial production system of large chemical and cement plants that came into existence through the 1960s were established on turnkey basis by Western firms. Though not on a turnkey basis, a large steel mill erected in the early 1970s relied almost entirely on foreign assistance. Initial production capability to operate and maintain such process-oriented plants stemmed largely from extensive training by foreign suppliers before, during, and after the erection of the initial production processes.

These firms, however, used deliberate strategies with considerable efforts to acquire capability not only for the operation and maintenance of processes but also the design and erection of new plants. Consequently, such efforts enabled the local firms to undertake a series of minor improvements, resulting in a significant productivity increase, and to progressively take charge of engineering tasks in the subsequent expansions. For example, POSCO, the only integrated steel mill in Korea, relied completely on the Japanese in the first phase in 1971 for annual capacity of one million tons. But by 1981 capacity had been increased to 8.5 million tons in three expansions that were increasingly under the direction of Korean engineers. The progressive substitution of foreign by local engineering is well indicated by the fall in the cost of foreign project engineering per ton from $6.13 to $0.13 during the same period. Some of the large firms with continuous process technology have accumulated experience and developed capabilities to export plants to other LDCs [12].

Another study on the pharmaceutical industry by Kim, Kim, and Lee [8], as presented in Table 4, shows two types of large firms classified by technological capability (first two columns in Table 4) exhibit distinctively different behavior in product development efforts. It shows, among other things, that large firms with high technological capability diversified their technological efforts, not only investing in their own efforts and in joint research with local R&D institutes, but also licensing foreign technology for their product development. Whereas, large firms with low technological capability resorted primarily to the transfer of foreign technology. Such a difference is also supported by a multivariate analysis of the same data (see [8] for results).

An interesting point in the pharmaceutical industry is that unlike other process-oriented industries, large firms in the pharmaceutical industry did not start from the beginning as a large scale firm. Rather, they evolved from two types of small firms to be mentioned below. In-depth historical data show that about a half of the large firms with high technological capability (type I) have evolved from a small scale with low capability (type IV) then through a large scale with low capability (type II), while others have evolved from a small scale with low capability (type IV) then through a small scale with high capability (type III). In other words, there are many indications that the four types of firms, as presented in Table 4, appear to show the growth path of local pharmaceutical firms in Korea. That is, many large pharmaceutical firms in Korea started as an importer/dealer of packaged finished drugs and later entered the drug manufacturing business by packaging imported bulk drugs. Then, they gradually extended into more intricate operations, first from formulating imported raw materials and later to backward integration into chemical com-

TABLE 4

Sources of technological inputs by two types of large and small pharmaceutical firms (Two-way ANOVA)[+]

Scale of operation Technological capability Variables	large high Type I (n=11)	large low Type II (n=9)	small high Type III (n=7)	small low Type IV (n=10)	main effect	firm size	tech. cap.	inter- action	ex- plained
Subjective data [++]									
Importance of local sources									
research institutes	3.36	0.89	3.86	1.00	***		***		***
suppliers	0.45	1.00	0.29	1.90	*		**		
competitors	0.27	1.00	0.71	3.70	***	***	***	***	***
government agencies	0.00	0.56	0.43	0.00				*	
individual consultancy	2.55	1.00	1.71	1.00			*		
technical manpower poached	1.18	1.00	0.00	2.60			*	**	**
Importance of foreign sources									
technical assistance	3.73	4.44	0.57	0.70	***	***			***
training program	3.55	4.22	0.00	0.00	***	***			***
suppliers	1.73	1.67	0.43	2.60				*	
publications	4.82	4.33	4.57	3.10	***	**	***		***
overseas observation	1.00	1.89	0.71	0.50		*			
Objective data									
Number of foreign assistance contracts	10.03	8.89	0.00	0.23	***	***			***
Number of technical licensing contracts	2.18	0.78	0.00	0.00	***	***	*	*	***
Number of contracts with local research institutes (75–79)	1.10	0.11	1.29	0.00	***		***		***
Number of contracts with local research institutes (80–84)	1.73	0.55	2.00	0.00	***		***		***

[+] Figures in the table indicate the mean value of each type.
[++] Subjective data are measured using 5 point Likert scale; 1, not important, through 5, very important.
*** significance level <0.01.
** significance level <0.05.
* significance level <0.1.
Source: Kim, Kim, and Lee [8].

ponents production. Through this process, they grew in size and in technological capability.

In contrast, Table 4 also shows that small firms with high technological capability rely mostly on their own R&D efforts with some assistance of local R&D institutes, whereas small firms with low capability imitate existing low-grade products by poaching experienced technical personnel from other firms. The data of two papermaking firms included in Amsden and Kim's study [4] and the observation of small chemical firms lead one to reach the same conclusion on technological behavior between firms with high capability and those with low capability as that reached in the study of small pharmaceutical firms mentioned above.

Though these firms were highly foreign dependent at the initial stage, they soon made deliberate efforts to participate actively in project execution and to assimilate imported technology related to plant design and operation. Because the final product and the equipment are relatively well known, the proprietary know-how that lies in the details of the production process can make a significant difference in the productivity of industries with process production technology. It was these deliberate efforts which made it possible for Korean firms to acquire technological capability so fast [9].

To recapitulate, large firms with process production technology also relied heavily on foreign sources for technological change. However, they exerted their efforts to localize technological contents in plant design and productivity improvement, while small firms relied on their own efforts to survive in a given market niche. Unlike other large firms using the continuous process technology, large pharmaceutical firms have progressively acquired technological capabilities, evolving from small, self-reliant to large modern firms.

5. A SYNTHESIS

A series of studies undertaken in Korean industries led to a tentative synthesis on the patterns of technological change in industries that use distinctively different operations technology. The skeletal description of the patterns, as presented above, is summarized in Table 5. Technological capability mentioned in the table may be subdivided into three major elements [13]: a production capability (ability to operate and maintain production processes), an investment capability (ability to design and erect new ventures and expansions), an innovation capability (ability to generate new products and processes and to improve existing ones).

There are several commonalities across three types of industries covered in this study. First, large firms across industries were highly foreign dependent in acquiring initial production technology. Why? Given the size of large investment and lack of technological capability and experience in early years, local firms were motivated to resort to well experienced foreign firms in order to ensure a quick erection and smooth start-up of their production processes and to acquire technological information and training to produce products with stringent specifications. Another reason may be that compared to small ones, large firms had organizational capability to identify and locate foreign technology and to negotiate with foreign technology suppliers, and financial resources to import foreign technology.

Second, large firms across the three types of industries exerted increasingly their own technological efforts to rapidly acquire technological capability and

TABLE 5

Summary of technological change in industries using different operations technology

		Source of initial technology	Indigenous efforts to assimilate foreign tech	Emphasis on acquisition of capability
Small batch and unit production	Large	Foreign licensings	High, rapid	Production* →Innovation*
	Small	Own imitative efforts	Low, slow	Production
Large batch and mass production	Large	Foreign licensings	High, rapid	Production →Investment →Innovation
	Small	n/a	n/a	
Process production	Large	Turnkey, consultancies	High, rapid	Production →Investment* →Innovation
	Small	Imitative	Low, slow	Production

*Production here refers to the capability to operate and maintain the production system. Investment refers to the capability to design and erect new ventures and expansions. Innovation refers to the capability to innovate and improve products and processes.

in turn reduce their reliance on foreign technology suppliers. There appear to be at least two reasons that account for local firm's motivation to augment foreign technology by their own efforts. First, perpetual dependency on foreign sources for new ventures and expansions is not only costly but also limits the economic activities of recipient firms through restrictive clauses in technology transfer contracts. Second, a series of minor improvements in product design and production operation through a firm's own efforts can result in significant improvement in competitiveness.

Third, all firms, large and small, placed emphasis on the acquisition of production capability at the outset in order to bring about immediate and efficient implementation of production operations. Those who used complex operations technology or those who produced sophisticated products relied primarily on foreign sources for help, while those who produced unsophisticated products with primitive operations technology were self-reliant.

Fourth, small firms across the industries took an imitative approach in developing the initial production processes and products, and evolved slowly for a long period of time in acquiring technological capability. The initial production processes were quite primitive ones developed by themselves after related foreign products. Why? First, the small firm had neither financial resources nor organizational capability to enter collaborative agreements with foreign suppliers. Second, it was relatively easier to adopt primitive operations technology for a small scale than for a large scale. Third, they catered to the market that used unsophisticated products.

There are also several differences across the three types of industries using different operations technology. First, though large firms across industries were commonly foreign dependent at the outset, the mode of their dependence varies according to the type of operations technology they deploy. For instance, industries with both small and large batch technologies resorted mostly to foreign licensings and consultancies. That industries with small batch technology were far more license-intensive than those with large batch technology may be accounted for by the fact that the former produces a larger number of highly differentiated products than the latter. In contrast, industries with process production technology initially imported turnkey plants. The difference between the first two and the last one stems from the fact that operations technology for process production is far more complex than the first two.

Second, all firms started with emphasis on the acquisition of production capability, but large firms soon diverted their efforts to acquire different types of capability so as to maximize the outcome from their efforts. For instance, large firms with small batch technology exerted their efforts to acquire innovation capability in order to modify and improve their products previously produced under licensing arrangements, because investment capability is less important for them; expansions may be done by adding more capitals. In industries with large batch operations, emphasis was shifted almost simultaneously to the acquisition of both investment and innovation capabilities after production capability, because ability to expand production system, which is more complex than small batch system, and ability to innovate new products were equally important. In contrast, those with process production went sequentially from the acquisition of production capability to that of investment and then to innovation. Because engineering costs were so expensive firms strove to internalize engineering capability in order to minimize investment costs. Then, on the basis of production and investment capabilities acquired, these firms went a step further to deepen their R&D efforts so as to innovate their processes and to undertake research in other areas.

Third, small firms resorted primarily to their own imitative efforts to develop their processes and products. But there were significant differences in their technological behavior in reverse engineering. For instance, the observation of related foreign products in use or for sale was the primary source of technological ideas for small firms in the machine industry, whereas the purchase of foreign specimens was the primary source for small firms in the electronics industry. The reason is that the cost of purchasing the related foreign product is beyond the financial capability of the former, while the latter can easily purchase foreign electronic products in the open market for reverse engineering purposes.

To recapitulate, there are both commonalities and differences in the pattern of technological change in industries with different operations technology and size. The similarities stemmed largely from the obsession of local firms to optimize learning given resources, while dissimilarities originated from differences in firm size and operations technology deployed by different industries.

6. DISCUSSIONS

Most studies on technological change are concentrated in developed countries, but the findings of these studies cannot necessarily be generalized to developing countries. Research of this type has just recently begun in newly industrializing countries which have increasing concern with technological change. This paper attempted to integrate the findings which emerged from extensive research undertaken in several industries using different types of operations technology in Korea.

Can the findings presented in this article be generalized to other countries? There are many fragmented pieces of evidence that support the patterns presented in this article. For instance, several studies of Japanese industrial history [14,15] offer evidence that Japanese industries evolved through a similar pattern in the past. American economic history also indicates that the U.S. reversed the sequence of research, development, and engineering (R,D,&E); that is, it started with practical technologies imported from European countries, and then slowly evolved into the present position of being able to afford substantial research [16]. Recent studies undertaken in other newly industrializing countries such as Argentina [17], Brazil [18,19], and India [20] also provide evidence to support the patterns described here.

However, it should be pointed out that the patterns presented here may not be generalized to those countries that have conditions different from those which Korea had. Korea had, among other things, an aggressive government policy that provided various incentives to induce import substitution productions. There were local entrepreneurs who know how to put foreign capital and technology and local human resources together to exploit new business opportunities created in the protected local market. There was also an abundant, highly skilled and motivated labor force, one that quickly learned and assimilated product design and production techniques. That is, ambitious investments in all levels of education in the early years provided an important base for the subsequent acquisition of technological capabilities for industrialization [21]. Thus, it is necessary to undertake similar research in other LDCs which have different conditions in key economic variables in order to develop more a generalized model of technological change in LDCs. The patterns presented here may serve as a departure point for generating a set of hypotheses that can be tested elsewhere.

NOTES

(1) It is the firm and a group of firms constituting an industry that bring about technological innovation. Both multinational companies that supply foreign technologies and local R&D institutions which help local firms import, adapt and improve such technologies interact primarily at the level of firms. Furthermore, the government uses a variety of programs within the general framework of their industrial policies in order to influence the process of technological change at the firm and industry level.

(2) Firms that deploy unit production technology are typically called job-order shops. Products are designed and manufactured on a "custom-made" basis. Consequently, they are relatively heterogeneous, and the tasks performed on each unit are typically noncomparable. Examples may include custom-tailored clothing, custom-made heavy machineries, ships, custom-made machine tools. At the opposite end of technological complexity, process production systems produce highly standardized outputs following a uniform and repetitive sequence. Production is characteristically maintained at a set level, and lot sizes are generally large. Illustration of this technology may include the manufacture of petroleum products, chemicals, and pharmaceuticals. Large batch systems fall between these two extremes. The operations performed in this type of technology are generally routine, repetitive, and predictable. Assembly line manufacturing of automobiles, large bakeries and mass produced clothing may be examples for this type of technology.

(3) It is characteristic of these firms that highly qualified "technical entrepreneurs" played a central role from the generation of the initial idea to the end of the problem solving in the innovation process.

(4) The main focus of Amsden and Kim [11] is the difference in technological and marketing behavior between a major local automobile producer and a joint-ventured subsidiary of General Motors.

REFERENCES

1 M. Goldsmith, (Ed.), Technological Innovation and the Economy. Wiley Interscience, London, 1970.

2 L. Kim, Stages of development of industrial technology a LDC: A model. Res. Policy, 9 (July 1980): 254–277.

276

3 J. Woodward, Industrial Organization: Thoery and Practice, Oxford University Press, London, 1965.
4 A.H. Amsden and Linsu Kim, Technology and Industrialization in Korea. Mimeograph, Prod. Div., the World Bank, 1985b.
5 A.H. Amsden and Linsu Kim, Technological perspective on the general machinery industry in the Republic of Korea. In: Martin Fransman (Ed.) Machinery and Economic Development, MacMillan Press, London, 1986, pp. 93-123.
6 L. Kim and Y. Kim, Innovation in a newly industrializing country: A multiple discriminant analysis. Manage. Sci., 31 (3) (March 1985): 312-322.
7 L. Kim and J.M. Utterback, The Evolution of Organizational Structure and Technology in a Developing Country. Manage. Sci., 29 (10) (October 1983): 1185-1197.
8 Y. Kim, L. Kim and J. Lee, Technological behavior of local pharmaceutical firms in Korea: A multivariate analysis. Working paper, Dept. Manage. Sci., Korea Advanced Institute of Science and Technology, 1985.
9 L. Kim, Imitation and apprentice: The acquisition of technological capability in Korea. College Bus. Admin., Korea University, Seoul, mimeo, 1987.
10 L. Kim, Technological innovation in the capital goods sector in Korea: A micro analysis. ILO working paper WEP 2-22/WP 92, International Labor Office, Geneva, 1982.
11 A.H. Amsden and Linsu Kim, The role of transnational corporations in the production and exports of the Korean automobile industry. Harvard Business School, working paper 90785-063, 1985a.
12 L.E. Westphal, Y.W. Rhee, L. Kim and A.H. Amsden, Exports of technology by the Republic of Korea. World Development, 12 (5/6) (May/June 1985): 505-533.
13 C.J. Dahlman and L.A. Westphal, Notes on the conceptualization of technology and technological capability. Dev. Res. Dept., the World Bank September, 1983. a mimeograph.
14 T. Ozawa, Japan's Technological Challenge to the West, 1950-1974: Motivation and Accomplishment. MIT Press, Cambridge, MA, 1976.
15 T. Shishido, Japanese policies for science and technology. unpublished mimeograph, Nikko Research Center, Tokyo, Japan, 1972.
16 NAS (National Academy of Sciences), U.S. International Firms and R,D,&E in Developing Countries. National Academy of Sciences, Washington, DC, 1973.
17 J. Katz, Domestic technology generation in LDCs: A review of research findings. Res. Prog. Sci. Technol. Dev. in Latin America, Buenos Aires, working paper no. 35, November 1980.
18 J. Baranson, The drive toward technological self-sufficiency in developing countries. Paper presented at the Conference on Latin America-United States Economic Interactions: Conflict, Accommodation, and Policies for the Future, University of Texas at Austin, March 19, 1973.
19 F. Sercovich, The acquisition of technological capability in Brazil. Dev. Res. Dept., the World Bank, 1985, a mimeograph.
20 S. Lall, The Acquisition of Technological Capability in India. Dev. Res. Dept., the World Bank, 1985, a mimeograph.
21 L. Kim, Technological transformation in Korea: Progress achieved and problems ahead. World Inst. Dev. Econ. Res., Helsinki, mimeo, 1986.

[6]

Pros and Cons of International Technology Transfer: A Developing Country's View

LINSU KIM

Recently, international technology transfer has been a subject of debate in both developed countries (DCs) and less developed countries (LDCs). Many argue that international technology transfer from DCs to LDCs should be promoted because it is mutually beneficial. They claim that international technology transfer to LDCs allows enterprises in DCs (1) to prolong the life cycle of products that are becoming obsolete in the home market, (2) to find new, growing markets, and (3) to ensure its own survival by relocating production segments to LDCs where labor costs are lower. Likewise, international technology transfer from DCs benefits enterprises in LDCs by enabling them to generate and improve products and processes and to gain export markets.

On the other hand, others argue that international technology transfer should be restricted to protect national interests; that its long-term effect is rather negative. Opponents in DCs argue that technology transfer to LDCs has "boomerang" effects, ultimately damaging their own industries and employment. Those opponents in LDCs argue that technology transfer results in an economic and technological dependence of LDCs on DCs.

This chapter assesses the pros and cons associated with international technology transfer from an LDC perspective. It presents a conceptual framework that may be useful to identify and assess different forms and channels of international technology transfer. With Korea as the case in point, it examines how an LDC acquires foreign technology from DCs and discusses implications for both DCs and LDCs.

A CONCEPTUAL FRAMEWORK TO ASSESS TECHNOLOGY TRANSFER

Foreign technology transfer may be mediated by the market; the supplier and the buyer may negotiate payment for technology transfer, either em-

Table 12.1 The Mode of Foreign Technology Transfer

Market mediated	Direct foreign investment, foreign licensing, turn-key plant, technical consultancy, made-to-order machinery (Cell 1)	Standard (serial) machinery purchase[a] (Cell 2)
Nonmarket mediated	Technical assistance by foreign buyers, technical assistance by foreign vendors[b] (Cell 4)	Imitation (reverse engineering) observation, trade journals, technical information service (Cell 3)
	Active	Passive
	The Role of Foreign Suppliers	

[a]Except for small, standard machinery, foreign suppliers send their engineers to assemble and test-run machinery sold. Often, they teach local personnel how to operate it and provide after-sale services. In this sense, the supplier's role is not passive, but compared with those mechanisms in cell 1, this mechanism can still be classified in cell 2.

[b]The vendor's service mentioned here refers to technical assistance not directly related to the operation of machinery sold; rather, the suppliers provide technical information and consultancy on operations not related to the machinery sold in exchange to a long-term purchase agreement.

bodied in or disembodied from the physical equipment. Foreign technology may also be transferred to local users without the mediation of the market. In this case the technology transfer usually takes place without formal agreements and payments. The foreign supplier can take either an active role, exercising significant control over the way in which technology is transferred to and used by the local recipient, or a passive role, having almost nothing to do with the way the user takes advantage of available technical know-how either embodied in or disembodied from the physical item. The recipient may be a producer, attempting to generate new products for import-substitution, or a user, deploying foreign technologies to improve productivity and product quality. These three variables—the mediation of the market, the role of foreign suppliers, and the role of recipients—offer a useful eight-cell matrix (Table 12.1) to identify and evaluate different mechanisms of international technology transfer and to assess its pros and cons.

Most studies of technology transfer have devoted their attention to the examination of those modes in cell 1 (market mediated with active suppliers). These include technology transfer accompanied by direct foreign investment (DFI), foreign licensing (FL), technical consultancy (TC), and made-to-order machinery. Recipients of foreign technology in this cell are mostly producers, with only a few exceptions.

Technology transfer in cell 2 is market mediated, but foreign suppliers play a relatively passive role, not exercising much control over the way in which technical know-how embodied in physical items is transferred to and used by the buyer. The purchase of standard serial machinery that embodies new technologies serves as the major means of transfer in cell 2. Recipients of foreign technology transfer in this cell are mostly users.

Technology transfers that take place in cell 3 are nonmarket mediated, in which the foreign suppliers play a relatively passive role. Reverse engineering done by local firms for their product and process development, reverse engineering by public research and development (R&D) centers for localization of technology beyond the capacity of local firms, and technical information services provided by various public agencies are good examples. Both producers and users could be in this cell.

In cell 4, technology transfer per se is not negotiated and priced in the market, but the foreign partners involved play a relatively active role in transferring technological know-how. There are at least two distinct cases. First, foreign buyers of locally produced goods under original equipment maker (OEM) arrangements deliberately provide technical know-how to ensure that local firms manufacture products according to the end buyer's stringent specifications. Second, foreign vendors of components and equipment provide local buyers with important technical services related to a system, for which they sell their products.

In conclusion, besides such formal mechanisms as direct foreign investment, foreign licensing, and technical consultancy, which have been studied at great length in the literature, technology may also be transferred across national boundaries through the various informal mechanisms outlined above. A discussion of how an LDC uses these mechanisms follows using Korea as the case in point.

TECHNOLOGY TRANSFER IN PRACTICE: THE CASE OF KOREA

Korea is one of many LDCs that has developed comprehensive policies and measures to promote, enhance, and manage technology flows from DCs (Baranson and Roark, 1985). First, this section presents a brief discussion of Korea's public policies related to foreign technology transfer. Then Korea's experience in technology transfer under this public-policy environment is discussed. The Korea case illuminates further discussion of the pros and cons of international technology transfer.

Technology Transfer in Cell 1

Direct Foreign Investment

The Korean government took initial steps to organize national policies for technology transfer in the 1960s. Korea's direct foreign investment (DFI) policy was quite free at that time, allowing any form of bona fide foreign capital, including fully owned subsidiaries, with extensive incentives. However, not much foreign investment came in during the 1960s, primarily due to questions about Korea's political stability and the uncertain economic outlook.

The government reversed its DFI policy in the 1970s, tightening controls, fearing that unlimited flow of foreign investment might adversely affect the domestic economy. Joint ventures received higher priority than wholly owned subsidiaries. A general guideline was adopted setting three criteria: eligibility, foreign ownership, and investment scale. Therefore, competition with domestic firms was seldom allowed in both domestic and international markets. Export requirements were forced on DFIs. Foreign participation ratios were basically limited to 50 percent, except for high technology and entirely export-oriented cases, rendering Korea one of a few countries with very restrictive foreign investment regulations.

In September 1980, the Korean government reversed its position, substantially liberalizing foreign investment guidelines. Another important reform was made effective from July 1984. The major change was a switch from the system of a "positive list" (listing which industries are open to DFI) to that of a "negative list" (listing those that are prohibited). The government's open policy for DFI is aimed at inducing the transfer of sophisticated new technologies and at promoting market competition for domestic firms, even within the domestic market, to intensify their innovation activities.

Under such a public policy environment, Korea has drawn an increasing amount of DFI through 1986 (Table 12.2), excluding the 1977–1981 period, which reflects an economic recession in Korea stemming partly from domestic political and social instability and partly from the global economic recession. Over 48 percent of the total DFI took place in the last period (1982–1986). Japan accounts for over 68 percent in the number of cases and 55 percent of the value, followed by the U.S. (20 and 26 percent, respectively). However, the average size of the American DFI is more than twice that of the Japanese. Chemicals, machinery, and electrical/electronics account for over 72 percent of the total DFI in the manufacturing sector and over 46 percent of the total DFI in Korea. Its proportion has steadily increased over time, as Korea has undergone a structural change from labor-intensive industries to relatively more knowledge-intensive industries.

The size of DFI and its proportion to total external borrowing are significantly lower in Korea compared with other newly industrializing countries (NICs) such as Taiwan, Singapore, and Brazil. One report (KEB, 1987) shows that Korea's stock of DFI in 1983 was only 7 percent of that in Brazil, 23 percent of that in Singapore, and less than half the size of that in Taiwan and Hong Kong. The proportion of DFI in Korea to the country's total external borrowing is only 6.1 percent, far below the levels of Singapore (91.9 percent), Taiwan (45 percent), and Brazil (21.8 percent). This figure appears to be the outcome of the nation's DFI strategy, under which Korea favored loans as a source of foreign capital and relied on other mechanisms to gain foreign technology.

Unlike other NICs, the contribution of DFI to economic development has not been significant in Korea. A recent study (Cha, 1983) shows that DFI's contribution to the growth of the gross national product (GNP) in Korea in the period from 1972 to 1980 amounted to only 1.3 percent, while its contribution to total and manufacturing value added was only

Table 12.2 Foreign Technology Transfer to Korea

(unit: U.S.$ million)

Source	1962–1966	1967–1971	1972–1976	1977–1981	1982–1986	Total
Japan	8.3	89.7	627.1	300.9	875.2	1,901.2
U.S.	25.0	95.3	135.0	235.7	581.6	1,072.6
Others	12.1	33.6	117.3	184.0	309.7	658.7
Total	47.4	218.6	879.4	720.6	1,766.5	3,632.5
Foreign licensing						
Japan	—	5.0	58.7	139.8	323.7	527.2
U.S.	0.6	7.8	21.3	159.2	602.7	791.6
Others	0.2	3.5	16.6	152.4	258.5	431.1
Total	0.8	16.3	96.5	451.4	1,184.9	1,749.9
Technical Consultancy						
Japan	—	12.1	7.7	20.8	89.2	129.8
U.S.	—	3.1	6.0	16.7	159.1	184.9
Others	—	1.6	4.8	17.2	84.0	107.6
Total	—	16.8	18.5	54.7	332.3	422.3
Capital Goods Imports						
Japan	148	1,292	4,423	14,269	20,986	41,118
U.S.	75	472	1,973	6,219	12,394	21,133
Others	93	777	2,445	7,490	53,338	64,143
Total	316	2,541	8,841	27,978	86,718	126,394

Sources: Ministry of Finance for data on direct foreign investment and foreign licensing, Ministry of Science and Technology for technical consultancy data, and Korea Society for Advancement of Machinery Industry for capital goods data.

1.1 percent and 4.8 percent, respectively, in 1971, and 4.5 percent and 14.2 percent, respectively, in 1980. Its contribution to employment was only 0.2 percent and 1.5 percent, respectively, for 1971 and 1980.

Foreign Licensing

Korea's policy on foreign licensing (FL) was quite restrictive in the 1960s. General guidelines stipulated that royalties should be set within 3 percent of sales and contract duration within 3 years for the manufacturing sector, and that no export restrictions were allowed. In the early years, such a restrictive policy on foreign licensings led some local licensees to bargain in purchasing generally known mature technologies. The restrictive guidelines were significantly relaxed in the 1970s to allow the transfer of more sophisticated technologies. The policy was further relaxed in 1984, opening for all industries and for all terms and conditions. The government also used public R&D centers to assist the private sector in identifying particular foreign technologies and their suppliers and in strengthening bargaining power in transfer negotiations.

Table 12.2 also presents basic statistics of FL to Korea. It shows that, as with DFI, over 67 percent of FL was signed in the last period (1982–1986) and that the United States leads Japan in the receipt of royalty payments from the late 1970s, despite half the number of cases. These numbers indicate that the average royalty per case is far greater for the United States than for Japan, reflecting a significantly higher degree of sophistication in technology licensed from the United States. Kim's (1984) microeconomic study of 70 Korea firms confirms that American technologies are more capital-intensive, sophisticated, and complex than Japanese technologies.

Technical Consultancy
Table 12.2 shows that compared with other modes of technology transfer, technical consultancy is not significant in terms of value. The last period (1982–1986) accounts for over 78 percent of the total, and the United States leads Japan in supplying engineering consultancy services to Korea. The U. S. lead is attributed to the sharp increase of its share in the last period.

Turnkey Plants and Machinery
While Korea was restrictive on DFI and FL through the 1970s, it relied heavily on turnkey plants and made-to-order machinery for technology transfer. Firms in the process-based industries established in the 1960s and early 1970s, such as chemicals, cement, steel, and paper, resorted to the turnkey mode. In these industries, investment size was relatively large and proprietary engineering know-how was critical, although equipment was generally known. For these reasons, the best alternative for local firms without engineering capabilities was to rely completely on experienced foreign firms, to minimize the risks involved in a large investment and the time required to achieve normal operations. However, no quantifiable data is available to assess the quantity of technology transfer associated with turnkey plants' transfers.

Technology Transfer in Cell 2

The Korean government played a crucial role in transferring foreign technology embodied in standard serial machinery. The rapid growth of the Korean economy has called for paramount investment in production facilities. For example, the index of gross domestic fixed capital formation has risen rapidly from 50.7 in 1953 to 662.7 in 1970 and to 3,546.2 in 1984, with 1960 set equal to 100. Capital formation as a percentage of gross national product has risen from 5.5 percent to 25.3 percent to 33.4 percent during the same period. However, government policy has been biased in favor of foreign capital goods as a way to strengthen the international competitiveness of capital goods user industries. Such a policy led to massive imports of foreign capital goods, transferring new foreign technologies embodied therein, at the cost of retarding the development of the local capital goods industry (Kim, 1987a). The import liberalization rate in the machinery industry was relatively high until the first half of 1971, giving capital-goods users almost free access to foreign capital goods. Moreover, the slight overvaluation of the local currency, the tariff exemptions on imported capital goods, and the financing of purchases by supplier's credits, which carried low interest rates relative to those on the domestic market, all worked to increase the attractiveness of capital goods imports (Rhee and Westphal, 1977).

Table 12.2 also present data on capital goods imports. Japan leads the United States and the other countries in supplying capital goods to Korea. The U.S. proportion has steadily increased over time, but its share remained about half of the Japanese share in the whole period. Furthermore, of the four categories of technology transfer listed in Table 12.2, capital goods imports far surpass other means of technology transfers in terms of value. Capital goods imports were worth 34 times the value of DFI, 72 times the value of FL, and almost 300 times the value of technical consultancies. In sum, the total value of capital goods imports is 21 times that of all other categories combined. The trend is similar for individual source countries. For example, the total value of capital goods imports from the United States is 10.3 times that of all other categories combined from the United States. Thus, Korea may have acquired more technology from DCs through the importation of capital goods than through any other means. Compared to NICs, such as Argentina, Brazil, India, and Mexico, Korea has relied more on the import of capital goods for foreign technology transfer than on DFI and FL (Westphal et al., 1985).

Technology Transfer in Cell 3

Technology transfers that take place in cell 3 are nonmarket mediated, in which foreign suppliers play a relatively passive role. Reverse engineering done by local firms for their product and process development is the most important mode in this cell; both producers and users take advantage of this mode. Reverse engineering by public R&D centers for localization of technology beyond the capacity of local firms and technical information services provided by various public agencies may also be good examples. Since technology transfers in this cell cannot be quantified, one has to be satisfied with fragmented pieces of evidence from empirical studies and other anecdotes.

Reverse Engineering by Private Firms

The Korean government also has measures to facilitate the transfer of foreign technologies through the mode in cell 3. For example, in 1979 the Korean government introduced a scheme to designate locally developed innovative products and offered preferential treatment, tax incentives, and 2-year protection from entry by foreign technology or local imitation as a way to foster local development efforts. In the case of the machinery industry, preferential financing is offered even to the purchaser of such products.

There appear to be four types of firms that undertake reverse engineering. The first type comprises producers. Kim's (1980) study of consumer electronics firms reveals that of 15 black/white TV assemblers in 1975, 11 entered the industry by reverse engineering done by experienced engineers poached from existing firms. A study of 42 innovative capital goods producers shows that these firms primarily deployed the forms and channels to technology transfer in cell 3 (see Kim and Kim, 1985). Bae and Lee's (1986) study of small and medium-sized firms in Korea also shows that reverse engineering was the major means of product development for these firms. A recent study on the transfer of factory automation (F) to Korea shows that the majority of 28 firms doing business in CAD/CAM, 11 in automatic warehousing, and 22 in robotics, used reverse engineering as a way to get FA technology (Kim, 1988a).

The second type of firm is users. As in von Hippel's (1976) concept of user-paradigm, FA users first developed equipment related to FA in Korea. As mentioned previously, they imported the first batch of FA equipment to improve their productivity, but they soon developed FA equipment by assimilating imported technology on the basis of their experience in using foreign equipment and their own R&D efforts to meet subsequent needs and needs within their *chaebol* (Korean version of Japanese zaibatsu) group. An electronics firm has reportedly developed 2,500 units of 45 different pieces of FA equipment ranging from part-inserting robotics and automatic printed circuit board inspectors to automatic wrapping and unmanned vehicles. Many chemical, cement, paper, and steel makers studied by Kim (1987b) largely resorted to turnkey plants for the initial setup, but used reverse engineering for subsequent expansions. All of these firms have progressively sequenced from the importation of foreign equipment and systems to the imitative production of their own models for import-substitution. These firms later became equipment producers to capitalize on their capabilities, as the local market slowly took shape.

The third type includes firms that entered their manufacturing business first as sales and service agents for foreign exporters. Such experience provided local agents with opportunities to assimilate foreign technologies. For example, most of the 28 firms in the CAD/CAM business started as local sales/after-service agents for foreign suppliers. Many of them are now investing for their own R&D for assimilation and import-substitution of foreign systems. A similar pattern is also observed in Korea's entry into the computer industry (Kim et al., 1987).

The fourth type of firm includes new technology-based small producers that are started by technical entrepreneurs who have recently spun-off from local universities, R&D centers, or existing firms. Given the initial capability in innovation, they began developing innovative equipment and systems through reverse engineering of sophisticated foreign products (see Kim, 1988b).

Reverse Engineering by Public R&D Centers
Public R&D centers, 16 in number in 1981 and then reduced to 8 by mergers, have played a major role in developing new technologies through reverse engineering. Public R&D centers undertake joint research with the private sector to help local firms acquire new technology without collaborating with foreign sources or to help them gain better bargaining power in technology transfer negotiations. An example of the former would be the Korean reaction to the Japanese refusal to license technology for polyester film production to Korea, in fear of losing its market in Korea; a public R&D center in collaboration with a local chemical firm successfully undertook a reverse engineering task to develop the technology. Korea now is the world's major supplier for audio and video cassette tapes.

Public R&D centers also play an important role in developing new technologies that have large economic externalities. For example, the Korean government invested about $300 million in developing design and production capabilities for semiconductors and computers. The government established a public R&D institute several years before the private sector's entry into computers and semiconductors in an attempt to gain first-stage experience in R&D in the new technologies and to generate experienced researchers and production workers. This R&D center has been the spearhead and backbone of R&D in computer and semiconductor design.

In short, the major role of public R&D centers in a developing country, as discussed by Utterback (1975), is to facilitate and lubricate foreign technology transfer by assisting in the private sector's acquisition of foreign technology, formally or informally; to solve immediate or short-term problems; and to undertake their own R&D activities to pave the road for the private sector to have an advantageous position at a later point in time in acquiring newer technologies.

Technical Information Service
The government also exerts efforts to facilitate the international transfer of scientific and technical (S&T) knowledge through its technical information services provided by various public agencies. Through several monthly periodicals and on-line computer systems, these agencies collect, process, and disseminate S&T information on a nonprofit basis.

Technology Transfer in Cell 4

Technology transfer takes place in cell 4 where technology transfer per se is not negotiated and priced in the market, but foreign partners play a relatively active role in transferring technological know-how. Like those in cell 3, qualitative assessments have to be made, because no quantifiable information is available for the modes in cell 4.

There are at least two distinct cases. First, foreign buyers of locally produced goods under OEM arrangements deliberately provide technical know-how to ensure that locally manufactured products meet buyers' stringent specifications. Second, foreign vendors of components and equipment provide local buyers with important technical services indirectly related to a system, for which they sell their products. The first mode of technology transfer, which was mentioned repeatedly by local firms in field interviews (e.g., Kim, 1987b; Westphal et al., 1981; Kim, 1980) is a by-product of the export-oriented industrialization strategy in Korea. Therefore, public policy for export promotion has had an important impact on foreign technology transfer.

In conclusion, the government has played an important role in managing and facilitating the transfer of foreign technology by regulating collaborative agreements concerning direct foreign investment and foreign licensing and by providing incentives and preferential financing to those who acquire foreign technologies through means other than collaborative agreements. The government has also contributed significantly to technology transfer through the development of an R&D infrastructure such as public R&D institutes and technical information centers, which have played an important role in helping local firms acquire better bargaining power in technology transfer negotiations and also in the reverse engineering of foreign technologies.

IMPLICATIONS FOR DEVELOPED COUNTRIES

The conceptual framework and empirical evidence in Korea lead to several questions that DCs need to address to gain a better understanding of the pros and cons of technology transfer to LDCs. Should DCs try to stop transferring technology to LDCs in fear of boomerang effects? If yes, then can DCs stop transferring technology to LDCs? The evidence presented above suggests that the answer to both questions is "no."

Any attempt to restrict technology transfer to LDCs through the modes in cell 1 will jeopardize economic growth and yet cannot prevent LDCs from obtaining technology. The restriction of DFI to LDCs will jeopardize the global strategy of multinational firms in DCs, while that of FL will shorten the economic life cycle of their technologies and products. If one DC refuses to transfer a technology, a sophisticated technology buyer in the LDCs can turn to an alternative source. Examples abound. When the Japanese refused to license peroxide technology to Korea in fear of losing its market in the country (Korea imported 100 percent of its peroxide from Japan), Koreans turned to America, who was willing to license the technology to enter the new market. When large semiconductor firms in Japan and the United States were reluctant to transfer semiconductor technology to Korea, Korean firms were able to negotiate a dozen licensing agreements with small semiconductor firms in the United States between 1983 and 1985. These small firms were willing to sell the technology for a quick infusion of cash during the business slump in semiconductors (Kim, 1987c).

Alternatively, a sophisticated technology buyer can turn to another mode of technology transfer. Examples are again plentiful. When the Japanese refused to license video cassette recorder (VCR) technology to Korea, Korean firms employed reverse engineering (the mode in cell 3) of Japanese products (the mode in cell 2), augmented by technical consultancy (a mode in cell 1) provided by foreign engineers recruited on a short-term basis by the Korean firms. Now Korea challenges Japan in the world market.

DCs cannot stop transferring technology to LDCs through the mode in cell 2, because the transfer of standardized machinery is merely a pure form of commodity trade. Any restriction on trade will hamper economic growth of the individual firms and, in turn, of the entire nation. The empirical data (the value of Korea's imports of capital goods was 21 times those of DFI, FL, and TC combined) and impressions gained in field interviews with Korean firms suggest that Koreans appear to have learned more from imported capital goods than from other modes of technology transfer. Kim and Kim's (1985) study offers useful insights as to how Koreans innovate by taking advantage of locally available foreign products.

DCs can do little to stop technology transfer through the modes in cell 3. Rather, it is the outcome of the absorptive capacity and entrepreneurship of firms and R&D centers in LDCs that take advantage of these modes. Foreign technologies transferred through other means serve as important inputs for reverse engineering. The most innovative and profitable papermaker in Korea, for example, never entered collaborative agreements with foreign partners in the form of either DFI or FL but progressively reverse-engineered foreign models in developing the most sophisticated, computerized paper-making line (Kim, 1987b).

Technology transfer modes in cell 4 are also important means of economic activity for multinationals from DCs. OEM manufacturing in LDCs is a strategy of international sourcing of components and end-products by firms in DCs to sustain price competitiveness in both domestic and international markets. Informal technical assistance provided by foreign component suppliers is simply an extra service needed to conclude sales and to enable successful implementation of their components by the buyer.

None of the technology transfer modes in the four cells can be restricted by DCs to reduce the boomerang effects from LDCs. All modes, except for those in cell 3, about which DCs can do little, are, in fact, vital economic means for the growth of firms in DCs.

A second question arises: If technology transfer to LDCs cannot and should not be restricted, then does technology transfer from DCs to LDCs directly result in boomerang effects in DCs? Data show otherwise. Figure 12.1 presents a market-mediated total technology transfer to Korea in dollar value from Japan and the United States, together with the two sets of trade data between Korea and these two countries. The figure also shows an inverse relation between technology transfer to and exports from Korea. The total value of technology transfer from Japan is almost double that from the United States. Korea's exports to the United States, however, are almost double those to Japan. Furthermore, in 1986, for example, the

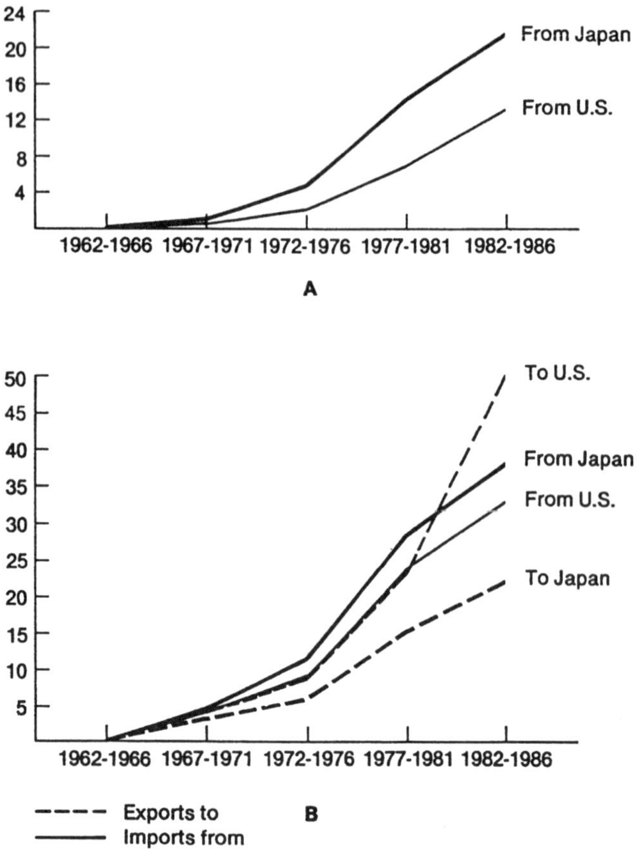

Figure 12.1 Technology and commodity trade between Korea and its two leading partners (1962–1986).

trade imbalance with the United States was almost $7 billion in favor of Korea, whereas that with Japan is $5.5 billion in favor of Japan. That is, Korea obtained foreign technology primarily from Japan but exported its products mainly to the United States. Therefore, Korea's export performance to the United States is not the direct outcome of the U. S. technology transfer to Korea. An Organization for Economic Cooperation and Development (OECD) study (1981) also concluded that trade competition came from countries that were not the biggest customers for technology and capital goods. Then, what is the source of the trade competitiveness of the fast growing LDCs? The international competitiveness of LDCs vis-à-vis a particular DC stemmed not so much from that DC's technology transfer to the country as from its absorptive capacity to assimilate, adapt, and improve foreign technologies. The internal weaknesses of DCs in industrial structure and innovation also account for the rapid export expansion of LDCs. Japan's postwar industrial history and the recent growth of Asian NICs (e.g., Taiwan, Korea, Hong Kong, and Singapore) demonstrate this point.

The third question emerges: What then is the net effect of the technology transfer of DCs to LDCs? Technology transfer definitely benefits recipient countries because it can be a vital source of technological change, leading to productivity improvement or to new products, processes and industries (see Kim, 1980). Technology transfer also benefits supplier countries. The OECD study (1981) concluded that the net return from technology transfers by DCs has been positive overall in terms of trade, employment, and consumer benefits in DCs. However, gross effects have fallen unequally on different industries. Engineering and machinery industries have benefited from increasing demand in LDCs during the period of low demand in other DCs. Some light manufacturing industries, however, had their problems compounded by the increased competition from LDCs. Kim's (1985) study of the United States–Korea link in technology transfer also concluded that although technology transfers from the United States to Korea have been accompanied by an increase in the volume of light commodity exports from Korea, its capital goods imports from the United States have increased dramatically to $6.2 billion in the 1977–81 period and doubled again to $12.4 billion in the 1982–1986 period. One might argue that the United States also faces increasing capital goods imports from Korea. American multinational corporations, however, comprise a significant part of the machinery industry in Korea. Helleiner (1979) shows that 64.2 percent of Korean general machinery exports and 67.3 percent of its electrical machinery exports to the United States were from American multinational corporations with production facilities in Korea.

Finally, what should DCs do to maintain technological leadership? No nation can maintain technological leadership through stringent controls of technology outflows: Witness the United Kingdom in the nineteenth century. Transfers are likely to continue at a significant rate, because on the supply side alternative sources of technology are increasing and the firms in possession of the technology may have to transfer it, as discussed previously, to expand sales and to extend the economic life of their technologies and products, maximizing the return from the technologies they possess. On the demand side, industries in some LDCs have developed increasing capabilities to master received technologies, to exert strong bargaining power in acquiring foreign technologies, and to undertake R&D to create their own innovations. Only through continuous innovation can DCs maintain their current position in technology leadership.

IMPLICATIONS FOR LESS DEVELOPED COUNTRIES

The conceptual framework and Korea's experience as presented above also offer useful implications for other LDCs that attempt to maximize benefits from technologies available in DCs. Three issues may be addressed in discussing the implications. The first issue is related to market-mediated technology transfers in cells 1 and 2. That is, LDCs should not view technology transfer from DCs to be a source of their foreign dependency and for this reason restrict it. The experiences of LDCs, like India, that adopted restrictive policies on foreign technology transfer and that stuck to policies of self-reliance, like the People's Republic of China in the 1960s and 1970s, have demonstrated that the preclusion of foreign technology can retard long-term economic growth. Rather, they should view technology transfer as a catalytic source of technological change, leading to the international competitiveness and economic growth of their countries. The experiences of other countries, like the Asian NICs, demonstrate that it is not technology transfer that leads to technological dependency on multinationals. Rather, it is the lack of local absorptive capacity to assimilate, adapt, and improve imported technologies that leads to dependency on foreign suppliers.

Nevertheless, heavy reliance on DFI as a means of technology transfer may, to some extent, lead to foreign dependency. DFI definitely transfers production capability (capability to operate and maintain a production system) but hardly investment capability (capability to set up or expand new production systems) or innovation capability (capability to innovate new products and processes), particularly when the parent company undertakes DFI to exploit the local market in LDCs. Kim's (1987c) study of microelectronics in Korea concluded that semiconductor manufacturing by multinational firms in Korea before 1984 only transferred simple packaging technology but fostered neither the skills, knowledge, and learning capabilities of the production workers nor the engineering capabilities of the domestic economy. Ernst (1985) reached a similar conclusion in his study of automation and the Third World. Amsden and Kim's (1985) comparative study of Hyundai and Daewoo (a joint venture with GM) in the automobile industry in Korea shows that the local firm outperformed the multinational subsidiary in product development and market performance, as the parent company constrained the latter's investment in R&D and GM was inactive in transferring technology transfer to its subsidiary.

The second issue is related to nonmarket-mediated technology transfer modes in cells 3 and 4. That is, LDCs can benefit greatly from informal technology transfers. It cannot be said in precise quantitative terms how important informal technology transfers have been in Korea, but earlier discussions, together with other studies (e.g., Kim, 1982; Kim and Kim, 1985) clearly indicate that they have been very important in Korea's acquisition of technological capabilities. This mode of technology transfer has clearly prevailed in innovative small firms and for a long time has been significant in broadening all exporters' capabilities (Westphal et al. 1985). In short, Korea's experience indicates that the majority of important or crucial information needed to solve technical problems can be obtained, free of charge, through nonmarket-mediated mechanisms, if LDCs have the local capability to undertake reverse engineering tasks.

The final issue concerns the views on foreign technology transfer and indigenous efforts in LDCs. Foreign technology transfer should not be viewed as a substitute for indigenous technological efforts or vice versa. Rather, they should be viewed as complementary. By providing new dimensions in technology, foreign technology transfer enables LDCs to make a quantum jump in indigenous technological efforts. Similarly, indigenous technological efforts lead to local ability to identify appropriate technology and to bargain better in technology transfer negotiations. Once imported, capabilities accumulated through indigenous efforts enable local firms to quickly assimilate, adapt, and improve transferred foreign technology.

SUMMARY AND CONCLUSIONS

The process of technology transfer is so diverse and complex that it defies a simple analysis. But the conceptal framework presented here enables one to identify various mechanisms of international technology transfer and to examine Korea's public policies promoting the inflow of foreign technologies and Korea's actual experience in the international transfer of foreign technology. The analysis led to the conclusion that informal mechanisms (i.e., those in cells 2, 3, and 4) are as important, if not more, as formal mechanisms (i.e., DFIs, FLs) in international technology transfer, particularly when the recipient country has absorptive capacity. This chapter then discussed the various implications of the conceptual framework and Korea's experience for both DCs and LDCs.

Table 12.3 schematically presents the discussions so far and their conclusions. In quadrant 2, where DCs transfer technologies but LDCs do not have absorptive capacity, DCs will enjoy their technological monopoly power and LDCs will suffer from the syndromes of technological dependency, a

Table 12.3 Assessment of Pros and Cons of International Technology Transfer

		The Existence of Absorptive Capacity in LDCs: Yes	The Existence of Absorptive Capacity in LDCs: No
Willingness of DC to transfer technologies through formal mechanisms to LDCs	Yes	Transfer takes place. Both DCs and LDCs gain (1)	Transfer takes place. DCs gain but LDCs become dependent (2)
	No	Transfer takes place. DCs lose but LDCs gain (3)	Transfer does not take place. Both DCs and LDCs gain little (4)

typical outcome of a zero-sum game. Neither DCs nor LDCs gain in quadrant 3, where DCs are not willing to transfer their technology LDCs and LDCs do not have the capability to benefit from existing technology in DCs. It is in quadrants 1 and 4, where LDCs have absorptive capacity, that DCs worry about the boomerang effects of technology transfer. But regardless of whether or not a particular DC is willing to transfer technologies through formal mechanisms, LDCs in quadrants 1 and 4 would be able to acquire foreign technologies, either through alternative sources or through alternative (informal) mechanisms. Why, then, do DCs not take advantage of opportunities to expand their market through DFI or to extend the economic life of technologies through FL (quadrant 1)? It is also in this quadrant 1 that LDCs will not end up being technologically dependent on DCs. That is, quadrant 1 is the best alternative for both DCs and LDCs. Here international technology transfer benefits both the supplier and recipient countries.

REFERENCES

Amsden, Alice H., and Kim, Linsu. (1985). "The role of transnational corporation in the Republic of Korea's production and exports of automobile," HBS Working Paper Cambridge, Mass.: Harvard Business School.

Bae, Zong-tae, and Lee, Jinjoo. (1986). "Technology development patterns of small and medium sized companies in the Korean machinery industry," *Technovation*, 4:279–296.

Baranson, Jack, and Roark, Robin. (1985). "Trends in north–south transfer of high technology," in Nathan Rosenberg and Claudio Frischtak (eds.), *International Technology Transfer: Concepts, Measures, and Comparisons*, New York: Praeger Press, pp. 24–42.

Cha, Dong-Se. (1983). *Weja Doipeo Hyogwa Boonsuk* (An analysis of the effects of direct foreign investment), Seoul: Korea Institute for Economics and Technology.

Ernst, Dieter. (1985). *Automation, Employment and the Third World - The Case of the Electronics Industry*, The Hague: IDPAD, Institute of Social Studies.

Helleiner, G. D. (1979). "Transnational corporations and trade structure," University of Toronto. Mimeo.

KEB (Korea Exchange Bank). (1987). "Direct foreign investment in Korea," *Monthly Review*, October, pp. 3–13.

Kim, Kee Young. (1984). "American technology and Korea's technological development" in Karl Moskowitz (ed.), *From Patron to Partner*, Lexington, Mass.: Lexington Book, D.C. Heath Co.

Kim, Linsu. (1980). "Stages of development of industrial technology in a LDC: A model," *Research Policy*, 9:254–277.

Kim, Linsu. (1982). "Technological innovation in the capital goods industry in Korea: A micro analysis," Working Paper, International Labor Office in Geneva.

Kim, Linsu. (1985). "Technology transfer and R&D in Korea: national policies and the U.S.–Korea Lin," *Korea's Economy*, 1(1): April.

Kim, Linsu. (1987a). *Technological Transformation in Korea: Progress Achieved and Problems Ahead*, a paper prepared for the World Institute for Development Economic Research, the United Nations University, Helsinki, Finland.

Kim, Linsu. (1987b). *Imitating and Apprentice: How Korea Acquired Technological Capability Fast*, College of Business Administration, Korea University.

Kim, Linsu. (1987c). *The Generation and Diffusion of Microelectronics: Local Capability, Employment Effects, and Public Policies in Korea*, a paper presented at a U.N. University New Technology Conference at the University of Limburg, Maastricht.

Kim, Linsu. (1988a). "The transfer of programmable automation to a rapidly developing country: An initial assessment," forthcoming in *International Economic Journal*.

Kim, Linsu. (1988b). "Entrepreneurship and innovation in a rapidly developing country," *Journal of Development Planning*, 18.

Kim, Linsu, and Kim, Youngbae. (1985). "Innovation in newly industrializing country: A multiple discriminant analysis," *Management Science*, 31(3):312–322.

Kim, Linsu, Lee, Jangwoo, and Lee, Jinjoo. (1987). "Korea's entry into the computer industry and its acquisition of technological capability," *Technovation*, 6:277–293.

OECD (Organization for Economic Cooperation and Development). (1981). *North/South Technology Transfer: The Adjustment Ahead*, Paris: OECD.

Rhee, Yung, W., and Westphal, Larry E. (September 1977). "A micro econometric investigation of choice of technology," *Journal of Development Economics*, 4:205–237.

Utterback, James M. (1975). "The role of applied research institutes in the transfer of technology in Latin America," *World Development*, 3(9):665–673.

Von Hippel, Eric. (1976). "The dominant role of users in the scientific instrument Iinovation process," *Research Policy*, 5 (3):212–239.

Westphal, Larry E., Kim, Linsu, and Dahlman, Carl J. (1985). "Reflections on the Republic of Korea's acquisition of technological capability" in Nathan Rosenberg and Claudio Frischtak (eds.) *International Technology Transfer: Concepts, Measures, and Comparisons*, New York: Praeger, pp. 167–221.

Westphal, Larry E., Rhee, Yung W., and Pursell, Garry (1981). "Korean industrial competence: Where it came from," World Bank Staff, Working Paper No. 469, Washington, D.C.: World Bank.

[7]

The Dynamics of Samsung's Technological Learning in Semiconductors

Linsu Kim

Despite the skeptics who claimed that it lacked the technological capability either to enter or remain competitive in the semiconductor industry, Samsung Electronics has in only a decade leapfrogged from a mere producer of discrete devices to the most vibrant producer of dynamic random access memory (DRAM) chips. Samsung has emerged as the world's largest memory chip maker and its seventh largest semiconductor make. Samsung's production increased from $83 million in 1985 to $5.2 billion in 1994 (see Table 1). As Business Week observed: "already dominant in 4-megabyte DRAM semiconductors, Korea is expected to stay ahead of Japan in 16-megabyte and 64-megabyte generations, while also attempting to crack more profitable applications-specific integrated circuits (ASICs)."[1] How has Samsung acquired the technological capability so expeditiously within a decade? What roles have institutions played in helping Samsung achieve phenomenal growth in technological learning?

Technological capability refers to the ability to make effective use of technological knowledge to assimilate, use, adapt, and change existing technologies. It also enables one to create new technologies and to develop new products and processes in response to the changing economic environment. The dynamic process of acquiring a technological capability is referred to as technological learning.[2] Technological learning is a function of firm's absorptive capacity, which has two important elements: a prior knowledge base and an intensity of effort.[3]

The case study underlying this article is based on Chapter 7 of the author's book, *Imitation to Innovation: The Dynamics of Korea's Technological Learning* (Harvard Business School Press, 1997).

CALIFORNIA MANAGEMENT REVIEW VOL. 39, NO. 3 SPRING 1997

TABLE 1. Production and Exports of Samsung's Semiconductor Business ($ million)

	1966	1970	1975	1980	1985	1990	1994
Production							
Korea Total	0.002	32	231	424	1,155	5,104	14,800
Samsung	—	—	0.4	14	83	1,188	5,246
Exports							
Korea Total	0.002	32	178	415	1,062	4,541	11,720
Samsung	—	—	0.4	7	75	1,130	5,000
Events	Multinationals entered assembly operations in Korea (1965)		Samsung began to fabricate wafers and produce LSIs (1975)	Samsung entered VLSI production under foreign license (1984)	Samsung became independent in DRAM design and production (1988)		Samsumg became the leader in DRAM design (1994)

Source: For Korea total data: Korea Development Bank, *Industry in Korea*, various years. For Samsung data: Samsung Electronics Co.

Prior knowledge base refers to existing individual units of knowledge available within the organization. Accumulated prior knowledge increases the ability to make sense of, assimilate, and use new knowledge. In countries that are catching-up, relevant prior knowledge base includes the basic skills and general knowledge necessary to support the relatively easy technological task as well as the most recent scientific and technological knowledge available from advanced countries.

Knowledge has two dimensions: explicit and tacit. Explicit knowledge refers to knowledge that is codified and transmittable in formal, systematic language. Explicit knowledge may be acquired in the form of books, technical specifications, and designs or as embodied in machines. Thus, explicit knowledge may easily be transferred through contractual or non-contractual mechanisms. In contrast, tacit knowledge refers to knowledge that is so deeply rooted in the human mind and body that it is hard to codify and communicate and can only be expressed through action, commitment, and involvement in a specific context. Tacit knowledge can be acquired only through experience such as observation, imitation, and practice. Thus, tacit knowledge can be transferred only through training or human transfer.

Prior knowledge at the firm is not a collection of explicit knowledge; rather, it is largely a collection of tacit knowledge. The firm may have some proprietary explicit knowledge, such as firm-specific blueprints and standard operating procedures. However, they are useful only when tacit knowledge enables its members to utilize them. Much of the knowledge that underlies effective performance in the organization is the tacit knowledge embodied in its members.[4]

The *intensity of effort* refers to the amount of energy expended by the organization's members to solve problems. It is insufficient merely to expose firms to relevant external knowledge without exerting effort to internalize it. Learning how to solve problems is usually built up over many practice trials on related problems. Thus, it requires considerable time and effort directed at solving simple problems early on before moving on to solving the more complex problems. Such effort intensifies interaction among the organization's members and facilitates knowledge creation and conversions at the organizational level.

Samsung's Entry as an Imitator

Korea's semiconductor industry traces its beginning to the mid-1960s, when several multinational semiconductor firms—Signetics, Fairchild, Motorola, Control Data, AMI, and Toshiba—began assembling discrete devices in Korea in order to take advantage of the local cheap labor. Operations involved simple packaging processes—bonded assembly operations by wholly owned foreign subsidiaries with all parts and components imported from the parent companies and exported back to the consignors. The assembly operations required only about six months training of unskilled workers and transferred little design or engineering capability to Korea.

As part of its heavy and chemical industries (HCI) drive, the government formulated a six-year plan in 1975 to promote the semiconductor industry. Many of the *chaebols*, the Korean version of the Japanese *zaibatsu*, showed great enthusiasm for entering the industry. However, unlike other HCIs, the difficulty in obtaining foreign technology and the high market risk associated with an increasingly short product life cycle chilled their zeal.[5]

It was in 1974 when the first local semiconductor firm was established by Dr. Ki-Dong Kang, a Korean-American scientist with a Ph.D. from Ohio State University and semiconductor design experience at Motorola. Samsung bought out the company during a financial crisis that occurred in the company's first year. With a large stake in consumer electronics, Samsung made the acquisition as a stepping stone to secure necessary prior knowledge related to semiconductors. Dr. Kang provided Samsung with an even higher tacit knowledge base, which was effectively transferred to Samsung engineers.[6] This enabled the firm to progressively produce various transistors and integrated circuits on a small scale for consumer electronics. Samsung also established the Semiconductor R&D Laboratory in 1982, which focused mainly on reverse-engineering and assimilating bipolar and metal oxide semiconductors (MOS).

The stakes grew substantially in 1983 when Samsung marshalled investment to enter very large scale integrated circuits (VLSI) production. The odds seemed to be stacked in its favor: it had ready access to funds syphoned from cash-cow industries within the *chaebol*, a motivated and highly skilled work force recruited from abroad, and a low labor-cost advantage. Leading foreign producers refused to license VLSI technology to Samsung, but Samsung was able to find a number of distressed small semiconductor companies in the U.S. that were eager to sell what Samsung needed most—chip designs and processes—in attempts to obtain cash for survival.

Heavy Reliance on Foreign Suppliers for the 64K DRAM

When it attempted to license 64K DRAM technology, Samsung was turned down by Texas Instruments and Motorola in the U.S. and NEC, Toshiba, and Hitachi in Japan.[7] Samsung then organized a task force team in 1982 to formulate an entry strategy for VLSI. Its eight years of experience in transistor and integrated circuit production provided a platform for VLSI. However, Samsung required a major technological leap from its current form of operations: from 5 micron to 2.5 micron in circuit width, from 3" wafer to 5" wafer,[8] and from 1K/16K large scale integrated circuits (LSI) to 64K VLSIs.

The task force team spent six months collecting information, analyzing the technology and market, and formulating plausible strategies. This significantly built up Samsung's tacit knowledge on VLSI. The team members then spent one month in the U.S. meeting experts in the industry, particularly Korean-American scientists and engineers working in American semiconductor firms or teaching at American universities. They studied the market and industry structure and identified potential technology suppliers. Their prior tacit knowledge enabled them to make sense out of discussions and observations during the trip.

Samsung then identified troubled small firms to buy VLSI technologies from. It licensed 64K DRAM design from Micron Technologies in Boise, Idaho, in a successful attempt to shorten the time in learning and production, gaining both explicit and tacit knowledge related to VLSI. Design for a high-speed MOS process was picked up for $2.1 million from Zytrex of California, which resulted in the transfer of explicit knowledge to Korea. Samsung sent its engineers to these technology suppliers for training as a part of the technology transfer process, elevating Samsung's capability to assimilate the licensed technologies.

Samsung sequenced the assimilation of transferred technology from the easiest to the progressively more sophisticated: from assembly processes to process development, then to wafer fabrication and inspection. Samsung first imported 3,000 64K DRAM chips from Micron Technology to assemble in Korea. With eight years of experience in the assembly of LSI chips, Samsung assimilated VLSI assembly technology without much problem, but this experience enabled Samsung engineers to gain initial familiarity with the new product. Samsung's assembly operation easily reached the 92 percent yield ratio, on a par with Japan. Then, Samsung went further to assimilate design and process technologies.

However, substantial technological capability and in-house research efforts were required to translate the knowledge acquired from the two American companies into a production operation with a high yield ratio. For this purpose, Samsung organized two R&D teams to work collaboratively in assimilating and commercializing the 64K DRAM.

Samsung set up one R&D outpost in Silicon Valley in 1983 and hired five Korean-Americans with Ph.D.s in electronics engineering from Stanford, Michigan, Minnesota, and Notre Dame Universities. These five also had semiconductor design experience at IBM, Honeywell, Zilog, Intel, and National Semiconductor. Along with some 300 American engineers, including several designers who left Mostek, these five scientists brought to Samsung the crucial tacit knowledge needed to crack VLSI technology. Silicon Valley was a strategic location for the development of the 64K DRAM. The high density of scientists and engineers in the vicinity offered the rich source of critical information and expertise that Samsung needed. The outpost also provided opportunities for engineers in Korea to participate in training and research in the U.S. and enabled them to learn about VLSI technology.

Samsung organized another task force team in Korea with two Korean-American scientists who had 64K DRAM development experience at American companies and Samsung engineers who were experienced in LSIs and trained on VLSIs at technology suppliers. Active interaction between the outpost in Silicon Valley and the team in Korea—through training, joint research, and consulting—effectively transferred knowledge from Silicon Valley to the Korean team in a very short period of time, making Samsung engineers better equipped to assimilate VLSI technologies from Micron Technology and Zytrex. In short, Samsung had a deliberate strategy to upgrade its prior tacit and explicit knowledge.

As a means to intensify its effort, Samsung created a crisis by insisting that the team develop a working production system for the 64K DRAM within six months. The team members lived together in make-shift quarters on the plant site and worked around the clock to assimilate imported technologies and develop processes. "I was so immersed in working on 64K DRAM that I stopped smoking and drinking. I hardly slept any more than three or four hours a day for six months," said Dr. Sang-Joon Lee, a Korean-American team leader.[9] The goal was clear to all team members. The determination to solve problems within the assigned time frame intensified close interaction among members. This led to a rapid knowledge conversion among the individual members and to a high rate of technological learning at the organizational level. Samsung managed the crisis as a creative, rather than destructive process. In short, a high prior knowledge base and high intensity of effort gave Samsung a high absorptive capacity and enabled it to rapidly acquire technological capability.

Beside the ad hoc meetings to solve spot problems, the task force team held regular meetings at 11 P.M. This meeting, called "eleven meeting," continued for several years. The "eleven meeting" later became the "nine meeting" and "seven meeting," as the sense of crisis dwindled down. After individuals and smaller teams worked all day, they regularly met during the night to discuss and coordinate progress made, problems faced, and plans for the next day. These meetings facilitated the creation and diffusion of knowledge among members and across smaller teams.[10]

Through their previous experience in semiconductor production, Samsung engineers had managed to assimilate and develop all but the eight core technologies needed for 309 processes in 64K DRAM production. After six months of hard work, the team in Korea assimilated the core technologies and succeeded in developing a "working good die."[11]

Next, Samsung set the ambitious goal of completing construction of its mass-production plant within six months, a task that in Japan normally required 18 months. This created another major crisis for Samsung engineers. The construction team collected and studied technical information to raise its relevant prior knowledge. It then identified a Japanese firm that had previously built a Sharp semiconductor plant in Japan. This firm designed the plant for Samsung and supervised its construction. This provided Samsung with an opportunity to learn about a Japanese system. Samsung switched its in-house assimilation efforts from Zytrex's untested technology to Sharp's production-tested technology. Samsung vitually copied Sharp's system to expedite its process development. Engineers from Samsung's process development teams and from its construction subsidiary worked around the clock including weekends and holidays. Despite a long series of trial and error, the team successfully completed the construction of a working plant within six months. Whenever problems arose, inexperienced Samsung engineers, not knowing much about the semiconductor mass-production process, had to check from A to Z to identify the source of problems. Such repeated checks of the whole system also helped Samsung accelerate its technological learning.[12]

Samsung hit the market with a 64K DRAM in early 1984, some forty months after the American pioneer and about eighteen months after the first Japanese version became commercially available. Korea became the third country in the world to introduce DRAM chips and significantly narrowed the technological gap with Japan and the United States.

The successful development of the 64K DRAM was the combined outcome of:

- high prior knowledge gained from new scientists and engineers recruited from the U.S. and licenses from small American firms,
- the high intensity of in-house efforts in assimilating and developing new DRAM products at the outpost team in Silicon Valley and the local team in Korea, and
- effective competition and collaborations between the two teams.

The Silicon Valley team took a leading role in this development process.

Dual Approach for the 256K DRAM

After meeting with success in the mass production of 64K DRAMs in early 1984, Samsung again launched two task force teams—one in Korea and the other at its outpost in Silicon Valley—for 256K DRAM development. Samsung used a dual approach, giving different assignments to the two task force teams. For the team in Korea, Samsung once again decided to source circuit design from Micron Technology to shorten the time gap with Japan and the U.S. in 256K DRAM commercialization. This time, however, it did not have to source process technology. Experience in developing the process for 64K DRAMs provided an invaluable platform for developing the process for the 256K DRAM. Nevertheless, several challenging process technologies for 256K DRAMs—the development of process for 2 micron circuits, 200 angstrom thin oxide fabrication, 1.1 micro meter metal pitch and chemical etching, test program and ceramic package assembly—posed another crisis on Samsung. The team went through all of the available literature on the 256K DRAM process and underwent intensive training at its supplier's site, again significantly raising the team's tacit knowledge base. Then, as the team did for the 64K DRAM, its members entered the crisis management mode, working around the clock for eight months. The team succeeded in developing a "working good die" in October 1984, reducing Samsung's pursuit of the world's pioneer from four years for the 64K DRAM to two years for the 256K DRAM. Mass production began in early 1986, about eighteen months after the first introduction by the advanced countries.

For the better-prepared outpost team in Silicon Valley, however, Samsung gave a different assignment. That team would develop the whole range of the 256K DRAM, including circuit design as well as process design, on its own in an attempt to become independent of foreign design suppliers. The team was to compete with its counterpart in Korea in developing process design. This created a major crisis in technological learning. The team spent seven months reverse-engineering 256K DRAM chips developed by Japanese and American firms and studying the literature on production processes. They completed the circuit design in April 1985 and a "working good die" in July 1985, about ten months after its counterpart in Korea.

The success of the Silicon Valley team had two important meanings. First, Samsung developed the technological capability to design a 256K DRAM circuit on its own, laying an invaluable platform for the subsequent development of the 1M DRAM. Through training and the relocation of personnel, design capability was effectively transferred to the Semiconductor R&D Center in Korea. Second, the quality of its working good die was better in terms of several important performance measures—such as soft error rate, electric static discharge, and information processing speed—than the model acquired from Micron Technologies. As a result, Samsung adopted the Silicon Valley version as the 256K DRAM design for mass production in Korea.

Facing the entry of Samsung and other Korean *chaebols* into the 64K DRAM and 256K DRAM markets, Japanese semiconductor producers moved quickly to dump their 64K and 256K DRAMs at a fraction of the Korean producers' cost. This strategy worked early on, placing enormous financial strains on their American competitors. However, unlike the single business semiconductor producers in the U.S., cushions provided by cash-cow subsidiaries within the diversified Samsung kept its semiconductor subsidiary afloat during the financial crisis. Then Samsung and other *chaebols* experienced a stroke of luck. Japan acceded to export restraints on semiconductor trade with the United States. This and the subsequent move to the 1M DRAM by Japanese firms opened up new opportunities for Samsung and other Korean firms to penetrate the U.S. market, allowing them to emerge as dominant suppliers of the 64K DRAM and 256K DRAM. Increasing demand and short supply also pushed prices for the 256K DRAM from $2 in 1986 to $5 in 1988. The market remained firm for many more years, leading Korean semiconductor producers into the black and firmly establishing them in the semiconductor industry.

In short, by the time the 256K DRAM was completed, the technological capability of the Korean team was now on a par with that of its counterpart in the Silicon Valley. Both had enough capability to undertake R&D on their own without foreign assistance.

Internal Competition and Collaboration for the 1M DRAM

In September 1985, when development of the mass production system for the 256K DRAM was at its final stage, Samsung's R&D team shifted its research focus to the development of the 1M DRAM. Despite the fact that Samsung could purchase designs from an American firm, it decided to go on its own.[13] Once again, the project was given to a team in Korea and a team at the Silicon Valley outpost. The project was in a sense competitive because they both were to work on the same project. It was in another sense collaborative because they were to exchange information, personnel, and research results.

Experience gained in the 256K DRAM development process provided a prior tacit knowledge base, but unlike the case of the 256K DRAM, Samsung had trouble securing explicit knowledge in the form of technical specifications, literature on production processes, and sample chips from pioneering firms in Japan and the United States. This preempted imitative reverse engineering.

Design and process development for the 1M DRAM required several significant technological changes from the 256K DRAM. They included, among other things, shifting design from N-MOS to energy-efficient C-MOS—which resulted in notable complications, such as circuit width change from 2 microns to 1.2 microns, from double-ploy process to triple-ploy process, and accelerating the processing speed by up to 100-120 nano seconds.

The task force team in Korea was headed by a Korean-American scientist and included five project teams: circuit design, unit process, devices, process structure, and test programs. Once again, research teams adopted the Korean style of crisis management, staying together in a make-shift housing arrangement and working around the clock. The team consulted extensively with scientists and engineers at universities and public research institutes in Korea and those abroad, including its counterparts in Silicon Valley. In March 1986, the team in Korea completed circuit design and by July 1986 it had produced a working good die, narrowing the gap with the Japanese pioneer from two years for the 256K DRAM to one year for the 1M DRAM. The outpost team in Silicon Valley also successfully completed the 1M DRAM development about three months later, indicating that the locus of R&D capability had by then shifted from the outpost to the R&D center in Korea.

In an attempt to narrow the gap in marketing 1M DRAMs, Samsung took the risk of building a mass-production system in parallel with R&D work. Samsung had gained substantial tacit and explicit knowledge in setting up mass-production systems for the 64K DRAM and 256K DRAM. This time Samsung designed and constructed production systems, needing only some technical consultations from Japanese and American firms. Samsung began mass-producing the 1M DRAM in late 1987, a year after Japanese firms, but in time to catch the rapid rise in market demand.

Samsung's Growth as an Innovator

However, the road ahead was getting bumpier. In 1986, Texas Instruments filed a suit against Samsung and eight Japanese chipmakers charging infringement of patents for DRAM designs, while Intel filed a similar suit against Hyundai and its American design suppliers. Both Samsung and Hyundai ended up paying royalties on the past and future sales of their memory products. Work on the next generation of chips—the 4M DRAM—meant competing neck-and-neck with Japanese and U.S. companies in exploring the frontiers of semiconductor technology. As the stakes rose in the chip game, the field of players grew smaller worldwide, meaning that few, if any, of those left in the game could be counted on to sell state-of-the-art chip design technology to the Korean *chaebols*. So, the Koreans had to tackle the 4M DRAM design alone.[14]

To cope with the new situation, Samsung intensified its R&D activities. Table 2 shows that its R&D investment increased from \$8.5 million in 1980 to \$891.6 million in 1994. Accordingly, the ratio of R&D to total sales also increased from 2.1 percent to 6.2 percent during the same period. The number of local patents granted also ascended drastically from four in 1980 to 1,413 in 1994, demonstrating significant achievements in new knowledge creation.

National Consortium for the 4M DRAM

The cost of semiconductor R&D grew exponentially with each succeeding generation. Anticipating difficulty in acquiring foreign technology and seeking to avoid costly duplication in research and investment, the government stepped in

TABLE 2. R&D activities at Samsung Electronics

	1975	1980	1985	1990	1994
Total Sales (W. 100 mil)	244	2,513	12,985	44,523	115,181
R&D Investment (W. 100 mil)	na	56	388	1,862	7,133
R&D/Total Sales (%)	na	2.1	3.0	4.2	6.2
R&D Personnel	na	690	1,821	6,686	8,919
Local Patents Applied For	na	18	309	1,732	2,802
Local Patents Granted	0	4	17	640	1,413
Foreign Patents Applied For	0		32	1,145	1,478
Foreign Patents Granted	0	0	2	128	752

Source: Samsung Electronics Company

and designated R&D on the 4M DRAM as a national project in October 1986. The Electronics and Telecommunications Research Institute (ETRI), a government research institute (GRI), served as a coordinator in the consortium of three *chaebol* semiconductor makers—Samsung, LG, and Hyundai—along with six universities. The objective was to develop and mass-produce the 4M DRAM by 1989 and completely close the technological gap with Japanese firms. The consortium spent $110 million for R&D over three years (1986-1989); the government shared 57 percent of the total R&D expenditure, a disproportionately large share in comparison to other national projects.

ETRI invited researchers from three *chaebols* to participate in developing core technologies jointly at ETRI facilities. The three *chaebols* were unwilling to work together, each forming their own working groups, trying to succeed independently. Consequently, the joint research program at ETRI had to be dissolved.[15] The *chaebols* took different approaches; Samsung worked on a stack structure and Hyundai on a trench structure, while LG worked on a hybrid structure.[16] Joint seminars organized by ETRI have, however, helped diffuse knowledge among the three *chaebols*. Samsung was the first among the three *chaebols* to announce its completion of designing a 4M DRAM in 1988, only six months after Japan. Samsung registered fifty-six patents related to the 4M DRAM compared to LG's forty, Hyundai's thirty-eight, the universities' three, and ETRI's eleven.[17] Samsung was able to mass produce 4M DRAMs at about the same time as the Japanese firms. Samsung had finally caught-up with the Japanese firms in memory chips.

Independent Approach beyond the 16M DRAM

The government also designated the development of the 64M DRAM and 256M DRAM as national R&D projects. Accordingly, a similar consortium was organized, involving ETRI as a coordinating GRI with three semiconductor producers. But the three *chaebols* had already built enough technological capability to work on their own and refused to share their knowledge with competitors. Samsung developed a 64M DRAM ahead of other competitors in August 1992. The three firms continued to invest in R&D to improve the 64M DRAM. Samsung, for instance, developed the second-generation 64M DRAM with a circuit width of 0.32 micron, 40 percent smaller than the first-generation, enabling the company to increase the productivity of an eight-inch diameter silicon wafer. In the second half of 1994, Samsung became the world's first supplier of commercial samples of 64M DRAMs to such giant users as Hewlett-Packard, IBM, and Sun.[18]

In August 1994, after investing $150 million in R&D over a thirty-month period, Samsung succeeded in developing the world's first fully working sample of 256M DRAM, made with 0.25 micron CMOS process technology. This was not only ahead of the government's schedule, but also ahead of Japan. Samsung also used its own patented technology to design a new architecture that facilitates input/output expansion and overcomes current operating speed limitations, making major improvements in the capacity to process vast amounts of data.[19]

Standing at the forefront of dynamic memory chip technology, Samsung now has something to offer to rival firms in Japan and the United States. As a result of these new opportunities, Samsung has entered into strategic alliances with rivals such as Toshiba, General Instruments, ISD, Mitsubishi, NEC, and Fujitsu. For instance, in September 1994 Samsung entered a strategic alliance agreement with ISD whereby Samsung helps them develop next-generation chips. At the same time, Samsung receives permission to use ISD's technologies on storage and reproduction of sound signals in semiconductors, which enables Samsung to develop chips with a storage capacity of up to fifteen times the existing models.[20]

Non-Memory Semiconductors

Samsung has emerged as a global contender in the memory chip business, but lags behind foreign competitors in non-memory devices, which constitute more than three quarters of the world's semiconductor chip market (about $101.3 billion in 1993). Samsung semiconductor has turned its eyes to the non-memory area with an ambitious vision of ranking among the world's top ten in this market within five years. Accumulated technological capability from its DRAM experience appears to provide the necessary platform for venturing into non-memory technology.

Samsung has moved swiftly into the non-memory chip business. Samsung has been aggressive in developing strategic ties with foreign firms, including Toshiba, ARM of Britain, and ISD of the United States. Samsung has also taken over HMS and IgT of the U.S. and has signed a technical partnership with American Array Microsystem, where Samsung holds a 20 percent equity stake. Rapid growth in the semiconductor market and Korea's dominance in memory chips has resulted in an enormous cash flow into Samsung, enabling it to make bold investment in in-house R&D and merger and acquisition in the non-memory business.[21]

Summary and Conclusion

In only a decade, Samsung leapfrogged over established competitors to become the leading memory chip producer. Samsung first licensed both design and process technologies from financially troubled small American firms for the 64K DRAM, but needed only design technology for the 256K DRAM. On the basis of these experiences, Samsung developed both design and process technologies for the 1M DRAM and subsequent chips on its own, rapidly displacing foreign assistance. In its own R&D efforts, Samsung established an R&D outpost in Silicon Valley and a semiconductor R&D center in Korea. In the beginning, the Silicon Valley outpost took the lead, while knowledge was transferred to Korean scientists and engineers through training, joint research, and transfer of personnel. As a result, the locus of technological capability shifted from the outpost to the center in Korea.

The process of Samsung's leapfrogging experience has several implications:

- First, international technology transfer takes place in many different modes. Technology may be transferred across national borders through formal mechanisms such as licenses. Technology may also be transferred through informal mechanisms such as literature, observation tours, sample products, and the mobility of personnel with new knowledge. The Samsung case shows that formal licensing played a central role in the initial stage, but informal mechanisms appear to have played even more important roles in the later stages as the technology recipient accumulated its own capability.
- Second, it is difficult for foreign firms to restrict the outflow of technology. When one source refuses to transfer technology, a sophisticated firm can secure an alternative source of technology, as was the case for the 64K DRAM. When large multinationals in Japan and the U.S. refused to license the technology to Samsung, it bought it from financially troubled small American firms. When firms refuse to supply technology, a sophisticated firm seeking the technology can develop its own technology, as was the case for the 4M DRAM. When Samsung had nowhere to turn, it entered a consortium with a local GRI and other local competitors to develop the technology on its own.
- Third, formal licenses may transfer knowledge to technology recipients, but it is their own tacit knowledge that enables them to assimilate and improve transferred knowledge. Knowledge may be transferred but the capability to use it cannot. For this reason, it is essential for a company

to build its own tacit knowledge before, during, and after technology is transferred in order to make most out of the transferred technology.

- Fourth, a mastery of technological learning at one stage provides a platform for the subsequent technological learning. Samsung's experience in consumer electronics provided a platform for it to enter the production of discrete semiconductor devices. This provided a platform to assimilate LSI technology, which in turn provided a platform for Samsung to assimilate VLSI technology. Tacit knowledge gained at one stage enabled Samsung to make sense of the explicit knowledge associated with the subsequent technology. It is, however, intensified in-house R&D that made it possible for Samsung to leapfrog in technological capability from each of these platforms.
- Fifth, prior knowledge base is one of the two important elements of absorptive capacity. There may be many different ways to build up a prior knowledge base. But the mobility of high calibre scientists and engineers appears to be the most effective way to leapfrog in the prior knowledge base. An overseas network of Korean scientists and engineers is expected to play a significant role in Korea's innovation drive in the future. All three Korean semiconductor producers—Samsung, Hyundai, and LG—relied heavily on the network of Korean-American scientists and engineers. But what made Samsung a pioneer in Korea is that it recruited more overseas-trained scientists and engineers and invested more heavily in R&D than other competitors.
- Sixth, another important element of absorptive capacity is the intensity of effort. Samsung uses crisis creation as a major means to intensify its effort, while goal consensus provides the direction for such effort. This makes the intensity of effort clearly focused on expeditious learning and knowledge accumulation. Crisis creation and expeditious learning are widespread in Korean manufacturing.[22] This process is most evident in Korean industries that are attempting to catch-up, including firms in consumer electronics, shipbuilding, steel, and machinery.
- Seventh, crisis creation may not be a useful tool for pioneering. In the case of catching-up, learning goals may be specific and clear and relevant knowledge is readily available elsewhere in various forms. Thus, the catching-up company can use crisis creation to generate enormous energy from organizational members in searching and converting knowledge at the individual and organizational levels. In contrast, the pioneering company has difficulty in identifying external sources of relevant knowledge and has to work with strategic ambiguity, which provides only a broad direction.[23] Consequently, learning may be creative but not necessarily expeditious.
- Finally, the government has formulated a series of promotional programs for the semiconductor industry. But unlike the case of other HCIs in

which the state played a major role in directing the *chaebols*, Korea's success in the semiconductor industry should be attributed to business initiative rather than state initiative.[24] The state appears to be in an inappropriate position to play a decisive role in highly dynamic sectors such as semiconductors. It is true that the government provided semiconductor producers with R&D subsidies in recent years, however the subsidies were granted not so much as the *chaebols* needed them as the government wanted to have a share in the success of memory chip development. The *chaebols* paid back the subsidies. The government also provided preferential loans to semiconductor producers, as it did to other designated industries. But its interest rate was far higher than that of foreign loans. Korean *chaebols* were in a strong position to generate the necessary financial resources through internal earnings as well as foreign credit and syndicated loans. Samsung mainly used these mechanisms to enter the highly capital-intensive VLSI production. GRIs have not proved to be flexible and dynamic enough to meet the needs of the private sector. In fact, *chaebols* are far more versatile and their R&D capability is more advanced than that of the GRIs in this highly dynamic sector.

Notes

1. See "Korea: Headed for High Tech's Top Tier," *Business Week*, July 31, 1995, pp. 32-38.
2. See Chapter 1 of Linsu Kim, *Imitation to Innovation: The Dynamics of Korea's Technological Learning* (Boston, MA: Harvard Business School Press, 1997).
3. See Wesley M. Cohen and Daniel A. Levinthal, "Absorptive Capacity: A New Perspective on Learning and Innovation," *Administrative Science Quarterly*, 35 (1990): 128-152.
4. See Richard R. Nelson and Sydney G. Winter, *An Evolutionary Theory of Economic Change* (Cambridge, MA: Harvard University Press, 1982), p. 134.
5. Changrok Suh, "The Political Economy of Competitiveness: The Case of the Korean Semiconductor Industry," a paper presented at the conference on Redefining Korean Competitiveness in an Age of Globalization, University of California at Berkeley, April 24, 1993.
6. Korea Semiconductor Company (KSC) was a joint venture with Integrated Circuits International, Inc. (ICII), an American company, where Dr. Kang worked previously. ICII's American engineers also played an important role in extending technical assistance to Samsung. Samsung later bought out ICII's equity share of KSC.
7. Jinjoo Lee, "*Urinara Giupeo Changeojuk Gisulgyebal Sunggongsarae*" [Successful Cases of Creative Technological Development in Korea], Korea Advanced Institute of Science and Technology, a mimeograph, October 1993, p. 5.
8. After attempting to assimilate untested 5" wafer production technology from Zytrex, Samsung switched to emulate a production-tested 4" wafer system used by Sharp.
9. Samsung Semiconductor and Telecommunications Co., *Samsung Bandoche Tongshin 10 Nyonsa* [Ten Year History of Samsung Semiconductor and Telecommunications] (Seoul: Samsung, 1987), p. 204.

10. See Chapter 6 of Youngrak Choi, *Dynamic Techno-Management Capability: The Case of Samsung Semiconductor Sector in Korea* unpublished Ph.D. dissertation at Roskilde University, August 1994, p. 106.
11. The working good die is a functionally operating wafer (at about the 99 percent level) before packaging and reliability test.
12. Youngrak Choi, op. cit., pp. 107, 137.
13. Samsung in fact acquired foreign technical assistance in various forms for solving minor problems and modifications. Youngrak Choi, op. cit., p. 110.
14. Linsu Kim, "National System of Industrial Innovation: Dynamics of Capability Building in Korea," in Richard R. Nelson, ed., *National Innovation Systems: A Comparative Analysis* (New York, NY: Oxford University Press, 1993), pp. 357-383.
15. Martin Bloom, *Technological Change in the Korean Electronics Industry* (Paris: OECD Development Center, 1992), p. 64.
16. The stack structure and trench structure are methods of increasing storage capacitors on wafers. The former stacks capacitors on wafers, while the latter etches wafers in a trench shape.
17. Kee-Young Kim and Ji-Dae Kim, "The Korean Electronics Industry," College of Commerce and Economics, Yonsei University, 1993, a mimeograph.
18. "Samsung Electronics Jumps Ahead of Competition in Memory Chip Output," *The Korea Economic Weekly*, January 16, 1995, p. 10.
19. "Samsung Announces First Ever 256M DRAM Chip Prototype Development" *The Korea Economic Weekly*, September 5, 1994, p. 8.
20. "Samsung Inks Semiconductor Tech Tie-Up Contract with US Companies," *The Korea Economic Weekly*, October 3, 1994, p.10.
21. "Non-Memory Capacity Expansion by Chip Makers Leads to Higher Sales," *Korea Economic Weekly*, August 14, 1995, p.9.
22. See Linsu Kim, "Crisis Construction and Organizational Learning: Capability Building in Catching-Up at Hyundai Motor," *Organization Science* (forthcoming, 1997); Linsu Kim, *Imitation to Innovation: The Dynamics of Korea's Technological Learning*, (Boston, MA: Harvard Business School Press, 1997).
23. Ikujiro Nonaka, "Creating Organizational Order Out of Chaos: Self-Renewal in Japanese Firms," *California Management Review*, 30/3 (Spring 1988): 57-73.
24. For an excellent discussion of the state role in the development of the semiconductor industry in Korea, see Jeong-Ro Yoon, "The State and Private Capital in Korea: The Political Economy of the Semiconductor Industry, 1965-1989," Ph.D. dissertation, Harvard University, 1989.

[8]

Crisis Construction and Organizational Learning: Capability Building in Catching-up at Hyundai Motor

Linsu Kim
Korea University, Seoul, Korea

Abstract

Effective organizational learning requires high absorptive capacity, which has two major elements: prior knowledge base and intensity of effort. Hyundai Motor Company, the most dynamic automobile producer in developing countries, pursued a strategy of independence in developing absorptive capacity. In its process of advancing from one phase to the next through the preparation for and acquisition, assimilation, and improvement of foreign technologies, Hyundai acquired migratory knowledge to expand its prior knowledge base and proactively constructed crises as a strategic means of intensifying its learning effort. Unlike externally evoked crises, proactively constructed internal crises present a clear performance gap, shift learning orientation from imitation to innovation, and increase the intensity of effort in organizational learning. Such crisis construction is an evocative and galvanizing device in the personal repertoires of proactive top managers. A similar process of opportunistic learning is also evident in other industries in Korea.

(*Organizational Learning*; *Absorptive Capacity*; *Crisis Construction*; *Knowledge*; *Catching-up*; *Hyundai Motor*; *Korea*)

Organizational learning and innovation have become crucially important subjects in management. Research on these subjects, however, is concentrated mainly in advanced countries (e.g., Argyris and Schon 1978, Dodgson 1993, Nonaka and Takeuchi 1995, Utterback 1994, von Hippel 1988). Despite the fact that many developing countries have made significant progress in industrial, educational, and technological development, research on learning, capability building, and innovation in those countries is scanty (e.g., Fransman and King 1984, Kim 1997, Kim and Kim 1985). Models that capture organizational learning and technological change in developing countries are essential to understand the dynamic process of capability building in catching-up in such countries and to extend the theories developed in advanced countries.

Understanding the catching-up process is also relevant and important to firms in advanced countries. Not all firms can be pioneers of novel breakthroughs, even in those countries. Most firms must invest in second-hand learning to remain competitive. Nevertheless, much less attention is paid to the imitative catching-up process than to the innovative pioneering process. For instance, ABI/Inform, a computerized business database, lists a total of 9,006 articles on the subject of innovation but only 145 on imitation (Schnaars 1994).

A crisis is usually regarded as an unpopular, largely negative phenomenon in management. It can, however, be an appropriate metaphor for strategic and technological transformation. Several observers postulate that constructing and then resolving organizational crises can be an effective means of opportunistic learning (e.g., Nonaka 1988, Pitt 1990, Schon 1967, Weick 1988), but no one has clearly linked the construct variable to corresponding empirical evidence.

The purpose of this article is to develop a model of organizational learning in an imitative catching-up process, and at the same time a model of crisis construction and organizational learning, by empirically analyzing the history of technological transformation at the Hyundai Motor Company (hereinafter Hyundai), the most dynamic automaker in developing countries, as a case in point.

Despite the prediction that none of South Korea's automakers will survive the global shakeout of the 1990s, having been driven out or relegated to niche markets dependent on alliances with leading foreign car producers (*Far Eastern Economic Review* 1992), Hyundai is determined to become a leading automaker on its own. Unlike most other automobile companies in developing countries, Hyundai followed an explicit policy of maintaining full ownership of all of its 45 subsidiaries, entering the

1047-7039/98/0904/0506/$05.00

auto industry in 1967 as a latecomer without foreign equity participation. Hyundai has progressed remarkably since then.

In quantitative terms, Hyundai increased its production more than tenfold every decade, from 614 cars in 1968, to 7,009 in 1973, to 103,888 in 1983, and to 1,134,611 in 1994, rapidly surpassing other automakers in Korea, and steadily ascending from being the sixteenth-largest producer in the world in 1991 to being the thirteenth-largest in 1994. Hyundai is now the largest automobile producer in a developing country. It produced its one-millionth car in January 1986, taking 18 years to reach that level of production in contrast to 29 years for Toyota and 43 years for Mazda (Hyun and Lee 1989).

In qualitative terms, Hyundai began assembling a Ford compact car on a knockdown basis in 1967. It rapidly assimilated foreign technology and developed sufficient capability to unveil its own designs, Accent and Avante, in 1994 and 1995, respectively. The company thus eliminated the royalty payment on the foreign license and was able to export production and design technology abroad.

Hyundai's rapid surge raises several research questions: (1) How did Hyundai acquire the technological capability to transform itself so expeditiously from imitative "learning by doing" to innovative "learning by research"? (2) How does learning in the catching-up process in a developing country differ from learning in the pioneering process in advanced countries? (3) Why is crisis construction an effective mechanism for organizational learning? (4) Can Hyundai's learning model be emulated by other catching-up firms? (5) What are the implications of Hyundai's model for future research? The following section briefly reviews theories related to organizational learning and knowledge creation. Then Hyundai is analyzed as a case in point to illustrate how the Korean firm has expedited organizational learning and to answer the research questions.

Crises and Organizational Learning

Organizational learning, whether to imitate or to innovate, takes place at two levels: the individual and organizational. The prime actors in the process of organizational learning are individuals within the firm. Organizational learning is not, however, the simple sum of individual learning (Hedberg 1981); rather, it is the process whereby knowledge is created, is distributed across the organization, is communicated among organization members, has consensual validity, and is integrated into the strategy and management of the organization (Duncan and Weiss 1978). Individual learning is therefore an indispensable condition for organizational learning but cannot be the sufficient condition. Organizations learn only when individual insights and skills become embodied in organizational routines, practices, and beliefs (Attewell 1992). Only effective organizations can translate individual learning into organizational learning (Hedberg 1980, Kim 1993, Shrivastava 1983).

Absorptive Capacity

Organizational learning is a function of an organization's absorptive capacity. Absorptive capacity requires learning capability and develops problem-solving skills. Learning capability is the capacity to assimilate knowledge (for imitation), whereas problem-solving skills represent a capacity to create new knowledge (for innovation).

Absorptive capacity has two important elements, prior knowledge base and intensity of effort (Cohen and Levinthal 1990). Prior knowledge base consists of individual units of knowledge available within the organization. Accumulated prior knowledge increases the ability to make sense of and to assimilate and use new knowledge. Relevant prior knowledge base comprises basic skills and general knowledge in the case of developing countries, but includes the most recent scientific and technological knowledge in the case of industrially advanced countries. Hence, prior knowledge base should be assessed in relation to task difficulty (Kim 1995).

Intensity of effort represents the amount of energy expended by organizational members to solve problems. Exposure of a firm to relevant external knowledge is insufficient unless an effort is made to internalize it. Learning how to solve problems is usually accomplished through many practice trials involving related problems (Harlow 1959). Hence, considerable time and effort must be directed to learning how to solve problems before complex problems can be addressed. Such effort intensifies interaction among organizational members, thus facilitating knowledge conversion and creation at the organizational level.

As shown in Figure 1, prior knowledge base and intensity of effort in the organization constitute a 2 × 2 matrix that indicates the level of absorptive capacity. When both are high (quadrant 1), absorptive capacity is high; when both are low (quadrant 4), absorptive capacity is low. Organizations with high prior knowledge in relation to task difficulty and low intensity of effort (quadrant 2) will gradually lose their absorptive capacity, moving rapidly down to quadrant 4, because their prior knowledge base will become obsolete as task-related technology moves along its trajectory. In contrast, organizations with low prior knowledge in relation to task difficulty and high intensity of effort (quadrant 3) will be able to acquire

Figure 1 **Dynamics of Absorptive Capacity**

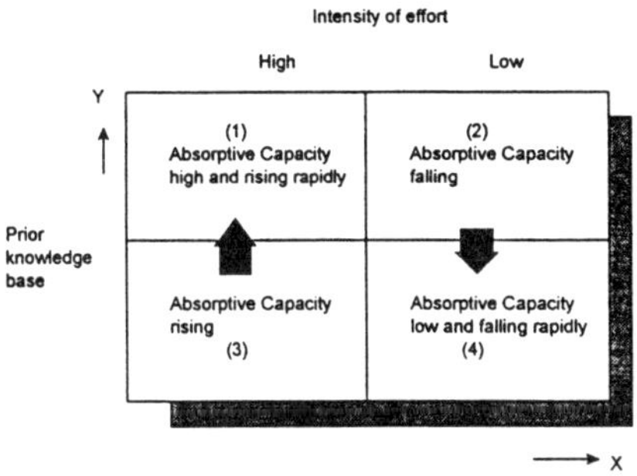

absorptive capacity, moving progressively to quadrant 1, as repeated efforts to learn and solve problems elevate the level of relevant prior knowledge (Kim 1995).

Knowledge and Learning

Many social scientists have attempted to delineate knowledge dimensions (Garud and Nayyar 1994, Kogut and Zander 1992, Polanyi 1966, Rogers 1983, Winter 1987). Polanyi's two dimensions, explicit and tacit, are the most widely accepted. Explicit knowledge is knowledge that is codified and transmittable in formal, systematic language. It therefore can be acquired in the form of books, technical specifications, and designs, or as embodied in machines. Tacit knowledge, in contrast, is so deeply rooted in the human mind and body that it is difficult to codify and communicate and can be expressed only through action, commitment, and involvement in a specific context. Tacit knowledge can be acquired only through experience such as observation, imitation, and practice.

Tacit knowledge is the core of a firm's prior knowledge base. The firm may have some proprietary explicit knowledge such as firm-specific blueprints and standard operating procedures. However, they are useful only when tacit knowledge enables its members to utilize them. Much of the knowledge that underlies the effective performance of an organization is tacit knowledge embodied in its members (Howells 1996, Nelson and Winter 1982).

Organizational learning takes place primarily through the dynamic process of four modes of conversion between the two dimensions of knowledge within the organization (Nonaka 1994, Nonaka and Takeuchi 1995). Tacit-to-tacit conversion (socialization) takes place when tacit knowledge within one individual is shared with another through training, whereas explicit-to-explicit conversion (combination) takes place when an individual combines discrete pieces of explicit knowledge into a new whole. Tacit-to-explicit conversion (externalization) takes place when an individual is able to articulate the foundations of his or her tacit knowledge, whereas explicit-to-tacit conversion (internalization) takes place when new explicit knowledge is shared throughout the firm and other members begin to use it to broaden, extend, and reframe their own tacit knowledge.

Figure 2 depicts the dynamic process of organizational learning in the catching-up process. It shows that prior knowledge base and intensity of effort affect the dynamics of knowledge conversion through a spiral process that starts at the individual level and moves up to the organizational level. Organizational learning tends to become faster and larger in scale as more actors in and around the firm with adequate prior knowledge intensify their efforts to convert knowledge within and between themselves. The outcome of knowledge conversion and creation feeds back to the prior knowledge base to increase its level.

In addition, migratory knowledge significantly affects the building of the prior knowledge base in the catching-up process (Badaracco 1991). Books, technical specifications, designs, and physical equipment transfer explicit knowledge, whereas the migration of individuals from one organization or country to another transfers tacit knowledge, elevating the level of the prior knowledge base.

Learning Systems

All organizations are learning systems. They learn as they develop, produce, and market products. All learning systems have a specific learning orientation reflecting the values and practices that determine what is learned and where. Learning orientation determines the way organizations acquire, share, and utilize knowledge. It might emphasize knowledge source, product-process focus, documentation mode, dissemination mode, learning focus, value-chain focus, or skill development focus (Nevis et al. 1995). In the catching-up process, as shown in Figure 2, different learning focus (duplicative imitation, creative imitation, or innovation) requires a different level of prior knowledge and a different degree of the intensity of effort. Learning orientation affects the spiral process of knowledge conversion.

Organizational factors (intention, autonomy, fluctuation and creative chaos, redundancy, requisite variety, and leadership) affect formal and informal processes and structures that facilitate organizational learning (Nonaka 1994, Nonaka and Takeuchi 1995). Redundancy and requisite variety reflect the characteristics of units of knowledge available in the organization. Redundancy in information and experience facilitates "learning by intrusion"

Figure 2 Dynamics of Organizational Learning in Catching-up

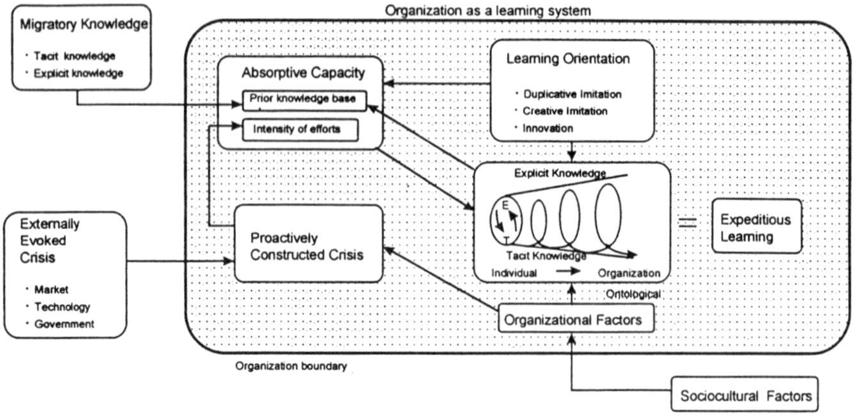

from different perspectives, whereas requisite variety advances the knowledge creation spiral by matching the variety and complexity of the environment. Intention and autonomy shape the knowledge frame that provides the ability to link units of knowledge and their priorities. Intention defines an organization's goals and fosters its employees' commitment, providing direction for the intensity of effort, and autonomy provides an environment in which a self-organizing team is able to function creatively. Fluctuation and creative chaos shape knowledge dynamics that foster the ability to manage the dynamic process in which individual units of knowledge are combined and transformed. Entrepreneurial leadership is also an important factor that creates organizational conditions conducive to learning.

As shown in Figure 2, the organizational factors also affect the spiral process of knowledge conversion and crisis construction. In addition, sociocultural factors influence the formation of work ethics. Vogel (1991) provides a seminal discussion of how cultural and situational factors formed work ethics in Korea and three other East Asian countries. What is most notable in Korean organizations is crises constructed proactively by top managers, which serve to intensify effort.

Crises and Learning

Cumulative or linear learning along the current trajectory can take place under normal circumstances. Discontinuous or nonlinear learning, however, takes place normally when a firm perceives a crisis and deploys strategy to resolve the critical situation (Meyers 1990). Organizations tend to engage in major changes mainly after they have been confronted with crises (Miller and Friesen 1984, Tushman et al. 1985). In such cases, the firm must invest heavily in the acquisition of new tacit and explicit knowledge as well as in knowledge conversion activities to overcome the crisis in the shortest possible time. The term "crisis" is expressed in Chinese using two characters, (weiji, 危機), the first meaning "danger" and the second "opportunity." Some firms manage to turn a crisis into an opportunity by transforming absorptive capacity in a discontinuous way to reap tremendous growth through enhanced competitiveness. A crisis may be creative in that sense; otherwise, it is apt to become destructive.

Crises may stem from external sources. They may be evoked naturally when the firm loses its competitive standing in the market and in technology. Literature abounds on market and technology-evoked crises (e.g., Abernathy 1978, Cooper and Schendel 1976, De Greene 1982, Meyers 1990, Miller and Friesen 1984, Shrivastava 1988, Tushman and Anderson 1986, Utterback and Kim 1985). A crisis may also be generated deliberately by an external principal. In developing countries, particularly where the state orchestrates industrialization, the government could impose a crisis by setting challenging goals for firms in a strategically designated industry. An external change generates a crisis for top managers but not necessarily for organizational members at the lower echelon.

Top managers can construct a crisis internally, either in response to or in the absence of an external crisis. A constructed crisis (Pitt 1990) may be set up deliberately at either the corporate level (corporate crisis) or the suborganization level (team crisis). Crises constructed at Hyundai are primarily team crises. Team crises may be more frequent and easier to manage than corporate crises, because they may have more focused and clearer goals. The shared sense of the internally constructed crisis among organizational members intensifies their efforts to expedite learning, elevating the absorptive capacity of the organization (see Figure 2). An organization with effective learning may frequently evoke constructed crises and institutionalize the process and structure to make discontinuous learning possible and turn crises into opportunities.

A case study was conducted to assess the impact of crisis construction on discontinuous learning in the catching-up process. It illustrates how the two elements of absorptive capacity, migratory knowledge, the four modes of knowledge conversion, organizational factors, externally evoked and internally constructed crises, and learning orientation affect organizational learning in a Korean firm.

Research Methodology

Stage 1

In 1984, an analysis was done to compare independent Hyundai and dependent Daewoo, a company engaged in a joint venture with General Motors. Company records, plant observations, and interviews with executives were used in the research (see Amsden and Kim 1989).

Stage 2

Subsequently, indepth case studies of Hyundai's technological learning were done almost once every two years. Again, information was obtained from Hyundai's company records, plant tours, and interviews with executives in manufacturing and R&D.

Stage 3

Hyundai published a well-documented history book (1,130 pages) in 1992 (see Hyundai Motor Company 1992), which provided rich historical information about the 1960s and 1970s and helped to verify and enrich case writeups of Hyundai's transformation in the 1980s and 1990s. It was supplemented with more recent records and interviews.

Stage 4

The case writeup covering 1967 to 1995 was completed in 1995. It was thoroughly reviewed by a senior engineer at Hyundai and an independent industry analyst.

Organizational Learning in Catching-up at Hyundai

Phase 1: Assimilation of Assembly Operations

Lacking experience in automobile production at the outset, Hyundai formed a taskforce in 1967. By drawing some team members from its construction division who had strong project management and engineering background, and recruiting some from other auto producers who had production experience, Hyundai ensured requisite variety in taskforce experience and knowledge. The recruited engineers brought in migratory knowledge that raised Hyundai's level of tacit knowledge related to automobile production.

In 1968 Hyundai entered into an Overseas Assembler Agreement with Ford, whereby Hyundai was to assemble Ford compact cars on a semi-knocked-down (SKD) basis. Ford transferred "packaged" technology to Hyundai with a set of explicit knowledge, such as blueprints, technical specifications, and production manuals. The agreement also included the training of Hyundai engineers at Ford sites and the dispatch of 10 Ford engineers to Hyundai to help translate the transferred explicit knowledge into tacit knowledge and to transfer to Hyundai Ford's tacit knowledge on procurement planning, procurement coordination, production engineering, process engineering, production management, welding, painting, after-service, and marketing. Also, Hyundai's suppliers sent their engineers to set up equipment and train Hyundai technicians. The most competent engineers trained by Ford were assigned to production and production engineering departments. In other words, the agreement with Ford gave Hyundai valuable migratory knowledge with which to upgrade its tacit and explicit knowledge related to auto assembly, moving the company up along the y-axis (toward a higher prior knowledge level) in Figure 1. Migratory knowledge alone is not sufficient, however. It must be reinvented by its user (Rice and Rogers 1980) through learning by doing (Arrow 1962) or learning by using (Rosenberg 1982).

At the same time, Hyundai constructed a crisis by setting an ambitious goal to complete plant construction in the shortest possible time to minimize lead time of production. Its engineers, technicians, and construction workers lived together in a makeshift structure on the plant site and worked 16 hours a day, seven days a week. The crisis generated intense interactions among the members, intensifying knowledge conversion in a spiral manner at the individual, group, and organizational levels and thus moving Hyundai leftward along the x-axis (toward higher intensity of effort) in Figure 1. Consequently, with high prior knowledge and high intensity of effort (quadrant 1), Hyundai achieved the shortest time (six months) between groundbreaking and first commercial production among the 118 Ford assembly plants around the world.

Hyundai also created a crisis for its production members by setting an ambitious goal to acquire a production capability in the shortest possible time. While plant construction was underway, production teams rehearsed production operations by disassembling and reassembling two passenger cars, a bus, and a truck over and over to routinize the production procedures, internalizing transferred explicit knowledge (production manuals) into tacit knowledge. When the plant was completed, workers had sufficient tacit knowledge to assemble cars with minimum trial and error. At the outset, technical emphasis was largely on the mastery of production capability to meet Ford's technical specifications. Rapid assimilation of production capability enabled the assembly production to evolve gradually from SKD toward a complete-knocked-down (CKD) operation. In short, in the first phase Hyundai expedited organizational learning by constructing an internal crisis in the absence of an external crisis.

Phase 2: Development of a "Korean" Car Under License

The second major jump in technological learning at Hyundai came in the mid-1970s, when the government imposed a crisis by making a radical policy change requiring the automobile industry to shift from assembly production of foreign cars on a CKD basis to the development of locally designed "Korean" cars. Policy implementation to develop "Korean" subcompact cars was highly centralized by the government, with the nation's president as the chief policymaker and the Ministry of Trade and Industry as a coordinating and implementing agency. In 1973, the government formulated The Long-Term Plan for Promotion of the Automobile Industry and ordered four automobile companies to submit detailed plans to develop Korean cars. Progress reports had to be briefed to the president regularly. The government's plan was very specific. For instance, the indigenous model had to be original, smaller than 1,500 cc in engine size with a local content ratio of at least 95%, less than $2,000 in production cost, and introduced in the market by 1975. The government also specified that production capacity per plant should be more than 50,000 units per year, at a time when the total annual passenger car production in the nation was a mere 12,751 units. To foster growth of the industry, the government established seven principles to promote indigenous subcompact model development. They included, among other things, protection of the local market from new entrants and from new foreign knock-down imports, a significant tax reduction, promotion of vertical integration leading to new business opportunities, preferential financing, tax concessions, and an administrative decree to guarantee a large market share for the indigenous model. Thus the crisis gave Hyundai an opportunity to expand its car business into the subcompact market under government protection.

In 1973 Hyundai submitted its masterplan for a new plant with a capacity of 80,000 "Korean" cars. Its actual production in that year was 5,426 cars. The plan represented a drastic departure from the past strategy of merely assembling foreign cars. It required the development of a highly successful "Korean" subcompact car that could be exported in substantial volume and simultaneously increase local market share enough to absorb the proposed production capacity. The plan posed a major constructed crisis for Hyundai engineers.

Although it lacked absorptive capacity, Hyundai decided to obtain foreign technologies from many different sources in an "unpackaged" form to maintain independence from foreign multinationals. However, the company had the clear goal of assimilating the imported foreign technology as rapidly as possible. Prior knowledge accumulated from mere assembly of largely foreign parts and components was inadequate for the new task. As the first step, Hyundai organized a project team and had its members master literature related to the various aspects of auto design and manufacture, thus accumulating new tacit knowledge converted from explicit literature-knowledge to enhance its prior knowledge level.

Hyundai approached 26 firms in five countries to acquire different technologies: 10 firms in Japan and Italy for car design, 4 firms in Japan and the United States for stamping shop equipment, 5 firms in the United Kingdom and Germany for casting and forging plants, 2 firms in Japan and the United Kingdom for engines, and 5 U.S. and U.K. firms for an integrated parts/components plant. The foreign companies gave Hyundai engineers observation tours not only of their own sites but also of the leading automobile manufacturing plants that were using the suppliers' technology, enabling Hyundai engineers to relate knowledge converted from literature to actual physical operations. Through that process, Hyundai engineers gained significant insight to large-scale, modern automobile manufacturing systems. Hyundai then entered into a licensing agreement with Italdesign for body styling and design, and with Mitsubishi for gasoline engine, transmission, and rear axle designs, as well as casting technology. Engineers were sent to those technology suppliers for training.

Hyundai acquired more foreign technologies from a greater variety of sources than its competitors in Korea. Up to 1985, Hyundai had signed 54 licenses with foreign suppliers in contrast to 22 for Daewoo, 14 for Kia, and 9 for Ssangyong. Licensing sources included Japan (22), the United Kingdom (14), the United States (5), Italy (5), Germany (3), and others (5). Mitsubishi accounted for only half of the Japanese licenses, reflecting Hyundai's independence in acquiring foreign technologies (Hyun and Lee 1989).

How Hyundai assimilated style design technology is illustrative. Hyundai formed a team of five design engineers, had them study literature related to auto styling, and sent them to Italy to participate closely with Italdesign engineers in the design process. Hyundai gave the team the ambitious goal of assimilating all the styling technology from Italdesign so that they could undertake subsequent style designs on their own. For one and a half years, the engineers lived together in an apartment near Italdesign, kept a record of what they were learning during the day, and had group reviews every evening. Such intensive interaction among the team members resulted in a very rapid spiral process of knowledge conversion within the team, significantly expanding Hyundai's tacit and explicit knowledge in styling. The team engineers later became the core of the style design department at Hyundai.

Although many engineers at Hyundai acquired necessary knowledge related to specific technologies, the company did not have experienced engineers who could put the knowledge together. To minimize trial and error, Hyundai hired for a three-year period (1974–1977) a former managing director of British Leyland as its vice president and six other British technical experts for the successful development of its first indigenous model, further increasing the prior tacit knowledge level of the firm. The technical experts, as the chief engineers of the chassis design, body design, development and testing, die and tooling, body production, and commercial vehicle design departments, played crucial roles in helping Hyundai engineers convert explicit knowledge supplied by licensors into tacit knowledge and integrate specific tacit knowledge into a workable system. After the British engineers left, Hyundai used moonlighting Japanese engineers to troubleshoot problems.

Hyundai developed its first indigenous model, Pony, with a 90% domestic content in 1975, making Korea the second nation in Asia to have its own domestic automobile. Hyundai quickly improved the car's quality over the years. It exported 62,592 cars to Europe, the Middle East, and Asia, accounting for 67% of Korea's total auto exports in 1976–1980 and 97% in 1983–1986. Pony accounted for 98% of Hyundai's exports during those periods. The company's local market share in passenger cars also increased, from 19.2% in 1970, to 59.4% in 1976, to 73.9% in 1979.

Crisis Construction and Shift of Learning Orientation. A constructed crisis speeded plant construction and expedited the assimilation of production technology. However, the continuing assembly operation (first vertical line in Figure 3) had a low performance goal and hence required little effort, leading to boredom and limited learning (point A in Figure 3), to a point where efforts had some property of decreasing return. The vertical direction of each succeeding phase has the same problem of decreasing returns on efforts.

Crisis construction shifted learning orientations (horizontal directions in Figure 3). The second constructed crisis shifted learning orientation from duplicative assembly of Ford cars on a CKD basis to the development of a "Korean car," moving to point (C). It provided a new learning frontier and increased return on effort. Because changing learning paths requires a huge amount of human psychological energy, top managers had to construct crises. Hyundai has continuously shifted its learning path—from a duplicative imitation-oriented one, to a more creative imitation-oriented one, to an innovation-oriented one—and in the process has accelerated its learning in a stepwise pattern, as shown in Figure 3.

Another important point is that Hyundai could not enter at point (B) from the beginning because of its insufficient prior knowledge base. Point (B) required too high a performance goal, given the company's prior knowledge base, and in turn too much efforts. Effort alone cannot expedite learning. It must be matched with the necessary prior knowledge to elevate absorptive capacity. Prior knowledge gained in the assembly of foreign cars provided a platform for the development of a Korean car.

Learning Process in Catching Up. A joint venture with a leading automobile firm is apt to lead to a passive attitude on the part of the recipient in the learning process, as technical assistance is always available from the parent company and the performance of the transferred technology is guaranteed by the supplier. In contrast, Hyundai pursued an independence strategy. It unpackaged technology transfer and independently took the responsibility of organizing imported technologies and components from multiple sources into a workable mass production system. That approach entailed a major risk because Hyundai would have to take total blame, not the foreign suppliers, if technology failed. However, it forced and motivated Hyundai engineers to assimilate foreign technologies as rapidly as possible throughout the process.

Each of Hyundai's learning phases apparently was associated with four stages: preparation, acquisition, assimilation, and improvement/application, leading to the rapid

Figure 3 **Crisis Construction and the Shift of Learning Orientation**

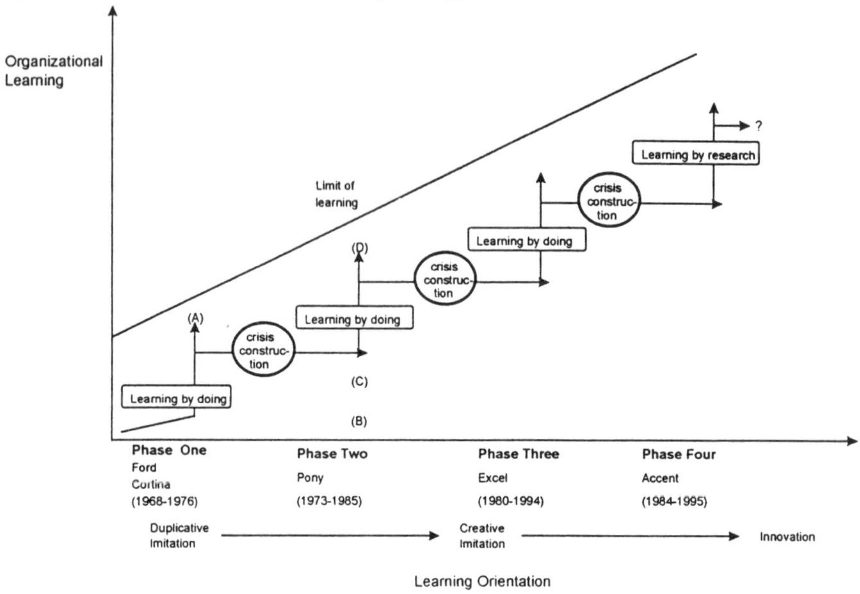

import substitution of personnel, engineering, and research in each phase (see Table 1).

At the outset of each learning jump, Hyundai took *preparatory* measures to acquire the migratory knowledge needed to elevate to its prior knowledge base. Those measures included poaching of experienced personnel from outside, extensive literature search, observation of technology in operation, and temporary hiring of foreign engineers. Experienced personnel from other firms or from abroad significantly raised the tacit knowledge base not only by the knowledge embodied in them, but also through knowledge socialization (from poached personnel to Hyundai personnel) prior to investment. The poached personnel constituted the core of Hyundai's task-force team.

Such an elevated prior knowledge base, albeit insufficient, enabled the team members to comprehend and internalize advanced explicit knowledge in literature. The team members were then sent abroad to observe facilities in operation in advanced countries, which gave them the opportunity to relate their knowledge from literature to actual physical settings in operation, further advancing the prior knowledge base. Whenever its tacit knowledge base was judged to be insufficient, Hyundai hired foreign engineers to augment its personnel.

The tacit knowledge base developed during the preparation phase helped Hyundai identify sources of foreign technology, strengthened its bargaining power in negotiations with foreign suppliers, and enabled it to *acquire* foreign technology on favorable terms. Foreign firms supplied both explicit (blueprints, technical specifications, equipment, manuals, etc.) and tacit knowledge (training and OJT supervised by foreign expatriates), but it was the prior knowledge base and intensity of effort at Hyundai that made the foreign technology transfer effective. Hyundai first relied on "packaged" technology transfer but rapidly unpackaged it as it developed the capability to substitute increasing portions of foreign input. Through preparing for and acquiring foreign technologies, Hyundai significantly increased its prior knowledge base.

Hyundai then intensified its in-house efforts to expedite the *assimilation* of imported foreign technologies. The

Table 1 **Spiral Process of Organizational Learning in Catching-up**[a]

	Phase 1	Phase 2	Phase 3	Phase 4
Cars produced	Ford Cortina	Pony	Excel	Accent
Technology mastered	Assembly technology	Initial design technology	Deepening design technology	Own design technology
Time period	1967–1976	1973–1985	1980–1994	1984–1995
Learning Stages				
Preparation	Poaching experienced personnel, literature review, observation tours	Literature review, observation tour, hiring foreign expatriates	Literature review, observation tour	Poaching scientists, literature review
Acquisition	Packaged technology transfer, hiring foreign expatriates	Unpackaged technology transfer	Unpackaged technology transfer	Acquisition by research, overseas R&D, hiring foreign expatriates
Assimilation	Learning by doing	Learning by doing	Learning by doing	Learning by research
Improvement/ Application	Learning by doing	Learning by doing	Learning by doing	Learning by research

[a]A constructed crisis facilitated expeditious learning at each phase. Prior knowledge base developed at the end of each phase provided a platform for the succeeding phase.

assimilation of know-how took place largely through "learning by doing" and "learning by using" in production. The assimilation of imported technologies led to incremental *improvement* of the imported technologies and their *application* to other areas. The sense of crisis intensified knowledge conversion at the individual level, and the shared sense of crisis heightened knowledge conversion at the organizational level through intense interaction among members.

After the first cycle of preparation to improvement/application in assembling Ford cars, Hyundai repeated a similar cycle in developing the Pony. Phases 3 and 4 also went through such a cycle. The mastery of one cycle provided the platform for starting the next cycle (Kim and Kogut 1995). Such repetition of similar cycles made up the spiral process of learning in catching-up.

Phase 3: Development of an Advanced Car Under Limited License

The third major jump in Hyundai's acquisition of absorptive capacity came in the early 1980s after the second oil crisis. In the face of losses due to rising gasoline prices and falling car sales, Hyundai decided to make a major investment to develop the next generation FF (front engine, front wheel drive) car to sell in the North American market, attempting once again to turn a crisis into an opportunity. The proposed production capacity was 300,000 units per year at a time when Hyundai was producing 57,054 passenger cars and the total number of cars produced in Korea was a mere 85,693 units. Hyundai was using only 32% of its production capacity (150,000 units). It was determined to turn the domestic-market-oriented automobile business into a largely export-oriented one. In response to the external oil crisis, Hyundai once again constructed a major internal crisis. The objective was to acquire the absorptive capacity needed to develop the next generation car and triple the production capacity.

Hyundai approached several major car producers—Volkswagen, Ford, Renault, and Alfa Romeo—for FF technology as a way to diversify its technology sources. However, all those companies demanded equity and management participation and viewed Hyundai as a local assembly subsidiary of their FF cars. Hyundai eventually approached Mitsubishi again. In 1981, Mitsubishi agreed to license engine, transaxle, chassis, and emission control technology to Hyundai. In return, Hyundai gave a 10% equity share to Mitsubishi without the latter's management participation. Not only did Hyundai retain all managerial control, but also reserved the right to import parts and technology from Mitsubishi's competitors and to compete directly in Mitsubishi's own markets. Hyundai sourced body styling from Italdesign and constant velocity joint technology from British GKN and Japanese NTN.

With the background of developing and manufacturing the Pony since 1976, Hyundai had a sufficient prior knowledge base to assimilate FF car design and manufacturing without foreign engineering assistance. However, technological learning was important in three areas.

The first was the manufacture of a car to meet the most stringent safety and environmental requirements of the United States. The second was the adoption of the computer-aided-design and computer-aided manufacturing (CAD/CAM) systems, the adoption of the assembly line control (ALC) system, and the development of transfer machines, which would lead to full computerization from design to manufacturing and parts/components handling and thus lay a crucially important foundation for Hyundai's development of cars on its own. The third was the construction of a proving ground, which would complete the infrastructure necessary for the next stage of strategy.

In the case of CAD/CAM systems, Hyundai organized a project team in March 1979 to develop a long-term plan to computerize design and manufacture. The project team collected literature and catalogs on CAD/CAM and spent the next 14 months internalizing explicit literature-knowledge into tacit knowledge. Using knowledge gained from literature, the team purchased an advanced computer system and a graphic plotter and undertook an in-depth study of Mitsubishi's operational CAD/CAM system. The team expanded to include two or three representatives from each department that would be affected by the CAD/CAM system, "socializing" tacit knowledge gained by the original members to new members. During the next 19 months, the expanded team determined the scope of CAD/CAM application and undertook an in-depth study of available alternative software packages. Hyundai selected the Catia package developed by French Dasso Aerospace in May 1982 and conducted preparatory work for almost 36 months before implementing the system.

Hyundai completed the FF plant in February 1985, tripling its production capacity from 150,000 units to 450,000 units per year. Its FF Excel passed both emission and safety tests and began being exported to the U.S. market in February 1986. Hyundai sold 168,882 Excels in 10 months during 1986, over 60% more than its ambitious goal, turning its loss to profit. With the sale of 263,610 in 1987, Excels became the best-selling import car of the year in the United States, overtaking Nissan Centra, Honda Civic, Subaru DL/GL, and Toyota Corolla.

Phase 4: Becoming Independent

The fourth jump in technological learning took place in September 1994, when the fiercely independent Hyundai unveiled Accent, the first subcompact car designed on its own (its predecessors had been based on Mitsubishi designs). It was benchmarked on the Toyota Tercel for performance and the Chrysler Neon for cost (Fortune 1995).

Although Hyundai was successful in manufacturing a subcompact FF car (Excel) and selling it in the North American market during the mid 1980s, it soon faced a technological dilemma. Mitsubishi, its major source of important technology for the FF car, and other foreign suppliers were unwilling to share their latest technology. Hyundai, however, lacked the absorptive capacity to keep upgrading its car quality to match that of its competitors in the North American market. Consequently, like Japanese export cars in the 1970s, Hyundai's Excel ended at the bottom of the *Consumer Reports* ranking, which greatly tarnished its image in the U.S. market. That experience prompted Hyundai to develop an extensive R&D network, not only in Korea but also abroad, to expand its absorptive capacity.

Hyundai's R&D efforts date back to 1978, when it established a primitive R&D center, primarily for facelifting current compact and subcompact cars. Efforts to develop its own capability, however, began to take shape in 1984 when Hyundai established the Advanced Engineering and Research Institute to develop its own engines and transmissions, the Passenger Vehicle R&D Center, the Commercial Vehicle R&D Center, and the Manufacturing Technology R&D Center. Hyundai also established joint R&D laboratories with local universities. In addition, in 1986 the Hyundai American Technical Center, Inc. (HATCI) opened in Ann Arbor, Michigan, and the Hyundai Styling Studio opened in Los Angeles to conduct R&D on cars for the U.S. market. Hyundai set up a technical center in Frankfurt to monitor technological developments in Europe and to design and engineer new cars for the European market. It opened an R&D center in Japan as the initial step toward entering that market in 1997.

The number of research engineers at Hyundai has increased, as shown in Table 2, from 197 in 1975 to 3,418 in 1990 and 3,890 in 1994, accounting for nearly 10% of the company's total employment; almost half of the engineers hold postgraduate engineering degrees. Hyundai recruited many Korean-American engineers from U.S. universities, some with experience at General Motors and Chrysler. For example, all but a few of the 35 senior research engineers with Ph.D. degrees at its Advanced Engineering and Research Institute are U.S. trained. Hyundai also invested heavily in continued training of its engineers. The number of R&D scientists and engineers sent abroad for training, ranging from short-term training and observation to long-term graduate degree programs, increased from 74 in 1982 to 351 by 1986. R&D investment also increased sharply, from 1.1 billion Korean won ($2.2 million) in 1975 to 400 billion won ($501.3 million) by 1994. R&D expenditures were 1.8% of Hyundai sales

Table 2 R&D Investment at Hyundai[a]

	1975	1978	1982	1984	1986	1988	1990	1992	1994
Sales[b]	30	216	430	669	1906	3411	4656	6079	9052
R&D[b]	1.1	5.4	7.9	22.7	79.5	116.0	190.4	248.8	400.0
R&D as % of sales	3.5	2.5	1.8	3.4	4.2	3.4	4.1	4.1	4.4
Number of researchers	197	381	725	1298	2247	2459	3418	3192	3890

[a]Hyundai Motor Company
[b]Unit: billion won

in 1982 and 4.4% in 1994, almost 60% higher than those of its domestic competitors such as Daewoo and Kia.

The development of its alpha engine illustrates how Hyundai became independent of Mitsubishi, which licensed Hyundai "old" engines but refused to share its state-of-the-art ones. As Hyundai lacked experience in designing even a carburetor engine, let alone an electronically controlled one, its decision to develop a state-of-the-art engine was another example of major crisis in technological learning. In 1984, Hyundai organized a taskforce with a vision of developing such an engine. However, none of the team members had any experience in engine design, and no car with an electronically controlled engine was available locally from which Hyundai engineers could learn.

The taskforce was divided into several teams to do research on (1) hydrodynamics, thermodynamics, fuel engineering, emission control, and lubrication; (2) kinetics and dynamics related to engine and car design and CAD; (3) vibration and noise; (4) new ceramics; (5) electronics and control systems; and (6) manufacturing control and CAM. More than 300 R&D personnel had received training overseas before the engine project was officially launched in 1984. Those team members collected all available English and Japanese language literature on engines and transmissions and mastered the literature to raise their prior knowledge level. Hyundai then entered into an agreement with British Ricardo Engineering, which provided initial assistance in technical training for engine design. Hyundai next hired two Korean experts who had gotten engine development experience at General Motors and Chrysler, respectively, after earning their doctoral degrees at American universities. Hyundai also hired an engineer from Ricardo for a three-year period beginning in 1985 (Hyun 1995).

Despite the training and consulting services by Ricardo and the three experts, Hyundai engineers underwent 14 months of trial and error before the first prototype was made. However, the engine block broke into pieces in its first test. New prototype engines were made almost every week, only to be broken again and again. No one on the team could figure out why the prototypes broke down during the test runs, and even Hyundai managers began to doubt the company's capability to develop a competitive engine. The team had to scrap 11 more broken prototypes before 1 survived the test. There were 288 engine design changes, 156 changes in 1986 alone (Hyun 1995). Ninety-seven test engines had to be made before Hyundai refined its NA (natural aspiration) and TC (turbo charge) engines, 53 more engines for durability improvement, 88 more for developing a car, 26 more for developing its transmission, and 60 more for other testing, totaling 324 test engines, as well as more than 200 transmissions and 150 vehicles before Hyundai perfected them in 1992.

Hyundai's alpha engine outperformed comparable Japanese ones. Hyundai's NA engine, for example, took 11.1 seconds to reach 100 km/h, whereas Honda CRX 3V took 11.30 seconds. Hyundai also outperformed Honda in fuel efficiency: 20.2 km/l versus 19.4 km/l for a four-gear manual transmission. The successful experience of developing the alpha engines (1.3 and 1.5 liters) led to the development of beta engines (1.6, 1.8, and 2.0 liters), making Hyundai completely independent of foreign license in engines for midsize, compact, and subcompact cars. Hyundai's new 1.8-liter, 16-valve, double overhead camshaft engine (DOHC), for instance, takes only nine seconds to reach 100 km/h, outperforming a similar model of Japan's Toyota (HMC 1992). With the two engines and its own transmissions, Hyundai reduced royalty payments for compact and subcompact cars to zero.

As a result of continuous R&D efforts, Hyundai achieved the largest domestic market share in 1995: its Accent is the best selling subcompact; the Avante is the best selling compact; the Sonata II is the best selling midsize car; and the Grandeur is the best selling full-size car in Korea. Hyundai also bounced back in its exports, particularly in new markets, from 225,393 units in 1990 to 392,239 in 1994. Hyundai is still behind the high-class Japanese and U.S. manufacturers, but the quality gap has definitely narrowed. The company's aim is to match Japanese quality but at a more competitive price. Hyundai has also become an important technology exporter to Thailand, Egypt, Zimbabwe, the Philippines, Malaysia, and other countries. It even exported a style design to Mitsubishi. Further, Hyundai is betting billions of dollars on a "green" car breakthrough. It plans to spend more than $5 billion on R&D from 1995 to 2000, lifting its R&D spending from 4.4% of sales in 1994 to 7.0% by 2000.

Behind the process of expeditious learning are Chung Ju-Yung, chairman of the Hyundai Group, and Chung Se-Young, his brother and chairman of Hyundai Motor. The group chairman is known as the most far-sighted and boldest risk-taking entrepreneur in Korea (see Kirk 1994). The history of his construction and shipbuilding divisions is also repeated with constructed crises and their successful transformation into creative learning (see Amsden 1989, Amsden and Kim 1985). He managed ongoing dialogues with taskforce leaders and their members. He visited sites regularly and maintained daily telephone contact with taskforce leaders to check the work in progress.

Summary and Discussion

Hyundai has transformed itself from a mere assembler of Ford models to a designer and exporter of its own cars and engines in less than three decades. Unlike firms in other developing countries, Hyundai has pursued a strategy of independence in developing technological capability. Its advancement from one phase to the next through the preparation for and acquisition, assimilation, and improvement of foreign technology appears to have been a spiral process. Migratory knowledge gave rise to prior knowledge base, and constructed crises intensified efforts, leading to expeditious organizational learning. The Hyundai case illustrates not only the process of expeditious learning, but also several idiosyncratic characteristics of organizational learning in catching-up.

First, catching-up firms, particularly those in developing countries, reverse the sequence of research, development and engineering (R.D,&E) of the advanced countries. Hyundai first acquired a production capability from its eight-year (1968–1976) experience in the assembly of Ford models. It then focused on the acquisition of an engineering capability, sequencing from subcompact to compact, to medium-size, to large cars. The number of years between steps steadily decreased: seven years (1976–1983) between subcompact and compact, five years (1983–1988) between compact and medium-size, and four years (1988–1992) between medium-size and large cars. It is logical for an inexperienced company in a catching-up country to sequence from a subcompact model, as its competitive edge depends more on price than quality in comparison with larger models. On the basis of production and engineering capability, Hyundai launched the development of its own cars. Again, Hyundai sequenced from subcompact (1994) to compact (1995), requiring only one year between the two. Hyundai is now doing serious research to break through in its "green" car project.

A similar process is evident in other industries in Korea. A series of studies covering more than 200 firms in different industries (Kim 1997) has shown that industries in Korea reversed the sequence of R,D,&E. They started with engineering (E) for products and processes imported from abroad in the 1960s and 1970s, then progressively evolved into the position of undertaking substantial development (D) in the 1980s. Research (R) became critical only in the 1990s.

Hyundai also reversed the sequence of technology trajectory of the advanced countries. Assembly technology from Ford in 1967 and technologies imported from various sources to produce the Pony in the mid-1970s were all mature technologies of advanced countries. Assimilating and improving those technologies through learning by doing enabled Hyundai to challenge more advanced technologies related to FF cars. Core technologies were again imported from advanced countries. The mastery of the FF technologies provided a platform for Hyundai's development of its state-of-the-art engines.

Second, migratory knowledge gives rise to a prior knowledge base. For catching-up firms, relevant knowledge is available elsewhere in various forms. The acquisition of prior knowledge through literature review and poaching of new personnel may be very effective for identifying and acquiring technology available elsewhere and facilitating learning in the subsequent phases. Particularly, mobile experienced personnel have been a major source of new tacit knowledge at Korean firms, as well as a major source of technology diffusion in many industries within Korea. The role of migratory knowledge and the government in the Hyundai case indicates the importance of institutions and interorganizational relations in organizational learning and innovation (Lundvall 1992). However, in-house efforts are indispensable to reinvent the transferred knowledge, particularly in the absence of mediating institutions that provide technology services (Attewell 1992).

Third, crises were constructed proactively rather than reactively. The first two crises (i.e., to expedite the assimilation of assembly production and to develop a "Korean car") were constructed proactively in the absence of external crises, and the third crisis (to develop an FF car and triple production capacity for the U.S. market) was constructed in response to the externally evoked oil crisis.

However, the external crisis served only as a trigger. Hyundai's internally constructed crisis was not designed to return Hyundai to its pre-oil-crisis position, but rather to transform Hyundai from a small automaker focusing primarily on the domestic market into an international contender challenging the most sophisticated U.S. market. It was proactive for long-term growth, rather than reactive to an externally evoked crisis for short-term survival. The fourth constructed crisis (to develop its own cars) was also proactive. The taskforce to develop Hyundai's own engines and transmissions was formed in 1984, four years before the company faced a market crisis in the United States, and most critical R&D problems were solved by 1986, two years before the market crisis. In other words, all the crises Hyundai constructed internally were proactive and at the suborganization level.

Why is an internally constructed proactive crisis, particularly at the suborganization level, more effective for organizational learning than an externally evoked crisis as a reactive means to respond to an external change? An evoked crisis creates a performance gap, a major discrepancy between how the organization performs now and how it ought to perform to survive. However, the phenomenon of crisis denial may occur: refusal to recognize that the crisis is real. It sometimes stems from different perceptions and interpretations of environmental changes among people involved and sometimes from active resistance to maintain the status quo or inertia in adhering to current norms and past practice. Consequently, the organization facing an evoked crisis must exert a significant portion of its energy to educate management coalition and organizational members to agree that there is a crisis, mitigate resistance to change, and unlearn past practices. Different perceptions of crises and diverse opinions on prescriptions also make it difficult for an organization facing an evoked crisis to direct its energy toward effective learning.

In contrast, constructed crises present a clear gap between the current performance and the performance needed in the future. Further, the top managers can manipulate the performance gap in constructed crises so as to make them creative rather than destructive. Constructed crises are an antidote to inertia. They generate intense pressure to create mandates for change (Meyers 1990), enabling the management coalition to reach consensus on organizational goals. Given unambiguous goals to close a performance gap, mandates for change, and determination of the management coalition, constructed crises also prompt members to accept organizational goals.

Crisis construction can be used as a strategic means to shift learning orientation. Hyundai used constructed crises to shift its learning from duplicative-imitation-oriented, to a more creative-imitation-oriented, and finally to innovation-oriented, continuously expanding the learning frontier to increase the return on effort. Because changing learning paths requires an enormous amount of psychological energy, entrepreneurial top managers had to construct crises.

Constructed crises also increase the intensity of effort at the individual and organizational levels in the search for alternative courses of action, and hence are creative rather than destructive (Kim 1997). Mandates for change generate volume of effort, whereas goal consensus and identification provide direction of effort, clearly focusing effort on expeditious learning for growth. The goal focus and high intensity of efforts to resolve crises not only prompt members to search actively for information on new ways to respond to them and to expedite knowledge conversion and accumulation at the individual level, but also intensify interaction among members, giving rise to knowledge conversion and accumulation at the organizational level.

Finally, past successes in transforming crises into creative learning evoke the self-confidence that leads to further risk-taking by crisis construction (March and Shapira 1987). A cycle from crisis construction to successful resolution to self-confidence and back to crisis construction characterizes Hyundai's history in accumulating an absorptive capacity.

Is Hyundai's learning model evident in other industries in Korea? A similar learning process is used, albeit to a different degree, in other Korean industries. For example, Korean firms in electronics (Kim 1980, 1997), shipbuilding (Amsden 1989, Amsden and Kim 1985), steel (Amsden 1989, Amsden and Kim 1985), machinery (Amsden and Kim 1986), and semiconductors (Kim 1997) have undergone a similar process of crisis construction and expeditious learning in catching-up. Crisis construction and expeditious learning are widespread in Korean manufacturing.

Hyundai is the most outstanding company in constructing crises proactively for opportunistic learning, mainly because it has the most entrepreneurial and boldest risk-taking top managers. In addition, the top and middle managers have been trained largely in the crisis construction and transformational process in Hyundai's project-based construction and shipbuilding companies. Those managers maintained ongoing dialogue during the process of crisis management, which created a "can-do" culture at Hyundai. Pitt (1990) concludes that crisis construction is an evocative and galvanizing device in the personal repertoires of proactive top managers, who think synthetically rather than analytically and are prophetic and single-minded.

Can Hyundai also use crisis construction for pioneering? Learning goals may be more specific and clearer in catching-up than in pioneering. The catching-up company can use crisis construction to achieve goal identification and consensus and to generate enormous energy from organizational members in searching for and converting knowledge at the individual and organizational levels. Relevant knowledge is readily available in various forms. The catching-up company can acquire prior knowledge through literature review, poaching of personnel, observation tours, and technology licensing. In contrast, the pioneering company must work with a strategic ambiguity that provides only broad direction (Nonaka 1988) and it has difficulty identifying external sources of relevant knowledge. Consequently, learning in pioneering may be creative but not necessarily expeditious.

Can catching-up firms in other countries emulate Hyundai's learning model? The answer is yes and no. They can improve the effectiveness of organizational learning by emulating the learning process illustrated in Figures 2 and 3 and the learning cycles listed in Table 1. But they may lack some of Hyundai's implicit advantages, such as well-developed human resources with the trait of being hardworking and cushions provided by the government and subsidiaries.

First, the availability of well-trained human resources cannot be easily matched by firms in other countries. According to a United Nations report, Korea is one of the only four developing countries that made a double jump from low level to medium level and from medium level to high level in terms of the human development index between 1960 and 1992. It had the largest absolute increase and the highest score among those four countries in 1992 (United Nations Development Programme 1994). The number of scientists and engineers per 10,000 population is the highest among the developing countries and closer to that of France and the United Kingdom, at least in quantity (Ministry of Science and Technology, Korea 1994). Human resource development requires a long-term investment.

Second, the Koreans' habit of hard work and their long working hours cannot be easily emulated by firms in other countries. Such cultural traits stem from several factors. Korea's land area is scarcely bigger than New Jersey in the United States, or Hungary. It is crossed by mountains, and a relatively small portion of the land is arable for its population of more than 44 million people. In population density, Korea trails only Bangladesh and Taiwan. The cramped conditions and severely cold winters appear to have forced Koreans to work hard and long to survive in an unfavorable environment. Additionally, the older generation has been motivated by the memory of deprivation and hard times under Japanese occupation and the destructive Korean War. An obsession to "beat Japan" to settle old scores and national economic competition with North Korea are also major forces motivating Koreans (Porter 1990). Such cultural and situational factors cannot be duplicated in other countries.

Third, government strategy and support, as well as cushions provided by cash-cow subsidiaries within the diversified Hyundai Group, enabled its automobile division to have a long-term focus and to take risks. For companies using equity market financing or operating in a single business, such long-term thinking and risk-taking may be difficult.

In conclusion, the Hyundai case leads to several speculative propositions to explore in the future. (1) A proactively constructed crisis is a more effective strategic means of shifting learning orientation and facilitating organizational learning than a reactively evoked crisis. (2) Entrepreneurial owners are more likely to construct and resolve crises than employed managers, particularly when they have governmental and interorganizational support for a long-term view. (3) A visionary entrepreneur, energetic teams, and effective dialogues between them are needed to turn constructed crises into expeditious learning. (4) Crisis construction and its transformation into expeditious learning may be easier to implement in catching-up than in pioneering. (5) Catching-up firms in both developing and advanced countries can elevate their prior knowledge base significantly by tapping migratory knowledge. (6) Catching-up firms in the developing countries reverse the sequence of technology trajectory of the advanced countries, entering first into the mature stage of technology and progressing gradually toward increasingly sophisticated technologies. (7) Catching-up firms in the developing countries sequence the process of capability building from production to engineering and innovation, reversing the direction of research, development and engineering (R,D,&E) common in the advanced countries. (8) For catching-up firms in developing countries, a strategy of independence is more difficult to manage but more effective in organizational learning than is joint venture with firms from advanced countries.

Acknowledgment

The author acknowledges the thorough and constructive comments from four anonymous reviewers, one of whom gave creative ideas underlying Figures 2 and 3. The case study underlying this article is based on Chapter 5 of the author's book, *Imitation to Innovation: The Dynamics of Korea's Technological Learning*, Boston, MA: Harvard Business School Press, 1997.

References

Abernathy, William J. (1978), *Productivity Dilemma*, Baltimore: Johns Hopkins University Press.

Amsden, Alice H. (1989), *Asia's Next Giant: South Korea and Late Industrialization*, New York: Oxford University Press.

—— and Linsu Kim (1985), "The Acquisition of Technological Capability in Korean Industries," World Bank, mimeograph (available from the authors upon request).

—— and —— (1986), "Technological Perspective on the General Machinery Industry in the Republic of Korea," in Martin Fransman (Ed.), *Machinery and Economic Development*, London: Macmillan Press, 93–123.

—— and —— (1989), "Comparative Analysis of Local and Transnational Corporations in the Korean Automobile Industry" in Dong-Ki Kim and Linsu Kim (Eds.), *Management Behind Industrialization: Readings in Korean Business*, Seoul: Korea University Press, 579–596.

Argyris, Chris and Donald A. Schon (1978), *Organizational Learning: A Theory of Action Perspective*, Reading, MA: Addison-Wesley.

Arrow, Kenneth (1962), "Economic Implications of Learning by Doing," *Review of Economic Studies*, 29, 166–170.

Attewell, Paul (1992), "Technology Diffusion and Organizational Learning: The Case of Business Computing," *Organization Science*, 3, 1, 1–19.

Badaracco, Joseph L. Jr. (1991), *The Knowledge Link: How Firms Compete Through Strategic Alliances*, Boston: Harvard Business School Press.

Cohen, Wesley M. and Daniel A. Levinthal (1990), "Absorptive Capacity: A New Perspective on Learning and Innovation," *Administrative Science Quarterly*, 35, 128–152.

Cooper, Arnold C. and Dan E. Schendel (1976), "Strategic Responses to Technological Threats," *Business Horizons*, 19, 61–69.

De Greene, K. B. (1982), *The Adaptive Organization: Anticipation and Management of Crisis*, New York: Wiley.

Dodgson, Mark (1993), "Organizational Learning: A Review of Some Literature," *Organization Studies*, 14, 3, 375–394.

Duncan, Robert B. and Andrew Weiss (1978), "Organizational Learning: Implications for Organizational Design" in B. Staw (Ed.) *Research in Organizational Behavior*, 1, 75–123.

Far Eastern Economic Review (*FEER*) (1992), "An Industry in Suspension," August 13, 25.

Fransman, Martin and Kenneth King (Eds.) (1984), *Technological Capability in the Third World*, London: Macmillan.

Fortune (1995) "Korea's Automakers Take on the World (Again)," March 6, 77.

Garud, R. and P. R. Nayyar (1994), "Transformative Capacity: Continual Structuring by Intertemporal Technology Transfer," *Strategic Management Journal*, 14, 365–385.

Harlow, H. F. (1959), "Learning Set and Error Factor Theory," in S. Koch (Ed.), *Psychology: A Study of Science*, New York: McGraw-Hill, 492–537.

Hedberg, Bo (1981), "How Organizations Learn and Unlearn," in P. C. Nystrom and W. H. Starbuck (Eds.), *Handbook of Organizational Design*, Vol. 1: *Adapting Organizations to Their Environments*, New York: Oxford University Press, 3–26.

Howells, Jeremy (1996), "Tacit Knowledge, Innovation and Technology Transfer," *Technology Analysis and Strategic Management*, 8, 2, 91–106.

Hyun, Young-Suk (1995), "The Road to Self-Reliance: New Product Development of Hyundai Motor Company" paper presented at International Motor Vehicle Program Annual Sponsors Meeting, Toronto, June 4–7, 1995 (available from the author upon request).

—— and Jinjoo Lee (1989), "Can Hyundai Go It Alone?" *Long Range Planning*, 22, 2, 63–69.

Hyundai Motor Company (HMC) (1992), *Hyundai Jadongcha Sa (The History of Hyundai Motor)* Seoul: Hyundai Motor Company, 1111.

Kim, Daniel H. (1993), "The Link Between Individual and Organizational Learning," *Sloan Management Review*, 35, 1, 37–50.

Kim, Dong-Jae and Bruce Kogut (1995), "Technological Platforms and Diversification" *Organization Science*, 7, 3, 283–301.

Kim, Linsu (1980), "Stages of Development of Industrial Technology in a Less Developed Country: A Model," *Research Policy*, 9, 3, 254–277.

—— (1995), "Absorptive Capacity and Industrial Growth: A Conceptual Framework and Korea's Experience," in Bon-Ho Koo and Dwight Perkins (Eds.), *Social Capability and Long-Term Economic Growth*, London: St. Martins, 266–287.

—— (1997), *Imitation to Innovation: The Dynamics of Korea's Technological Learning*, Boston: Harvard Business School Press.

—— and Youngbae Kim (1985), "Innovation in a Newly Industrializing Country: A Multiple Discriminant Analysis," *Management Science*, 31, 3, 312–322.

Kirk, Donald (1994), *Korean Dynasty: Hyundai and Chung Ju Yung*, Armonk, NY: M. E. Sharpe.

Kogut, Bruce and Udor Zander (1992), "Knowledge of the Firm and the Replication of Technology," *Organization Science*, 3, 383–397.

Lundvall, Bengt-Ake (Ed.) (1992), *National Systems of Innovation—Towards a Theory of Innovation and Interactive Learning*, London: Pinter Publishers.

March, James G. and Z. Shapira (1987), "Managerial Perspectives on Risk and Risk Taking," *Management Science*, 33, 1404–1418.

Meyers, Patricia W. (1990), "Non-linear Learning in Large Technological Firms: Period Four Implies Chaos," *Research Policy*, 19, 97–115.

Miller, Danny and Peter H. Friesen (1984), *Organization: A Quantum View*, Englewood Cliffs, NJ: Prentice-Hall.

Ministry of Science and Technology, Korea (MOST) (1994), *1994 Report on the Survey of Research and Development in Science and Technology*, Seoul: MOST, 75.

Nelson, Richard R. and Sidney G. Winter (1982), *An Evolutionary Theory of Economic Change*, Cambridge, MA: Harvard University Press, 134.

Nevis, Edwin C., Anthony J. DiBella, and Janet M. Gould (1995), "Understanding Organizations as Learning Systems," *Sloan Management Review*, Winter, 73–85.

Nonaka, Ikujiro (1988), "Creating Organizational Order Out of Chaos: Self-Renewal in Japanese Firms," *California Management Review*, 30, 3, 57–73.

—— (1994), "A Dynamic Theory of Organizational Knowledge Creation," *Organization Science* 5, 1, 14–37.

—— and Hirotaka Takeuchi (1995), *The Knowledge-Creating Company*, New York: Oxford University Press.

Pitt, Martyn (1990), "Crisis Modes of Strategic Transformation: A New Metaphor for Managing Technological Innovation," in R. Loveridge and M. Pitt (Eds.), *The Strategic Management of Technological Innovation*, Chichester, England: Wiley, 253–272.

Polanyi, M. (1966), *The Tacit Dimension*, London: Routledge and Kegan Paul.

Porter, Michael E. (1990), *The Competitive Advantage of Nations*, New York: The Free Press.

Rice, Ronald and Everett Rogers (1980), "Reinvention in the Innovation Process," *Knowledge: Creation, Diffusion, Utilization*, 1, 4, 499–514.

Rogers, Everett M. (1983), *Diffusion of Innovation*, New York: Free Press.

Rosenberg, Nathan (1982), *Inside the Black Box: Technology and Economics*, Cambridge, UK: Cambridge University Press.

Schnaars, Steven P. (1994), *Managing Imitation Strategies*, New York: Free Press.

Schon, Donald A. (1967) *Technology and Change: The New Heraclitus*, Oxford, UK: Pergamon.

Shrivastava, Paul (1983), "A Typology of Organizational Learning Systems," *Journal of Management Studies*, 20, 1–28.

—— (1988), "Industrial Crisis Management: Learning from Organizational Failures," *Journal of Management Studies*, 25, 4, 283–284.

Tushman, Michael L. and Philip A. Anderson (1986), "Technological Discontinuities and Organizational Environment," *Administrative Science Quarterly*, 31, 439–456.

——, B. Virany, and E. Romanelli (1985), "Executive Succession, Strategic Reorientations and Organizational Evolution," *Technology in Society*, 7, 297–314.

United Nations Development Programme (UNDP) (1994), *Human Resource Development Report, 1994*, New York: Oxford University Press.

Utterback, James M. (1994), *Mastering the Dynamics of Innovation: How Companies Can Seize Opportunities in the Face of Technological Change*, Boston: Harvard Business School Press.

—— and William J. Abernathy (1975), "A Dynamic Model of Process and Product Innovation," *Omega*, 3, 6, 639–656.

—— and Linsu Kim (1985), "Invasion of Stable Business by Radical Innovation," in Paul R. Kleindorfer (Ed.), *Management of Productivity and Technology in Manufacturing*, New York: Plenum Press, 113–151.

Vogel, Ezra F. (1991), *The Four Little Dragons: The Spread of Industrialization in East Asia*, Cambridge, MA: Harvard University Press.

von Hippel, Eric (1988), *The Sources of Innovation*, New York: Oxford University Press.

Weick, Karl (1988), "Enacted Sensemaking in Crisis Situations," *Journal of Management Studies*, 25, 4, 305–317.

Winter, Sidney G. (1987), "Knowledge and Competence as Strategic Assets," in David Teece (Ed.), *The Competitive Challenge—Strategies for Industrial Innovation and Renewal*, Cambridge, MA: Ballinger, 159–184.

Accepted by Ikujiro Nonaka; received February 1996. This paper has been with the author for one revision.

PART II

NATIONAL SYSTEM AND PUBLIC POLICY

[9]

Technology policy for industrialization: An integrative framework and Korea's experience *

Linsu Kim

College of Business Administration, Korea University, Seoul 136-701, South Korea

Carl J. Dahlman

Industrial Development Divisions, World Bank, Washington DC 20433, USA

Final version received August 1991

First, this article reviews three policy perspectives. (1) The market mechanism perspective includes the demand for and supply of technology and the linkage between the two. (2) The technology flow perspective consists of technology transfer, diffusion and R&D. (3) The dynamic perspective refers to technological trajectory. It then integrates these three perspectives to develop a new three-dimensional framework that may be useful to analyze technology policy for industrialization. Second, this article applies the framework to analyze Korea's technology policy as a case in point. Conclusions are: (1) Technology policies on foreign technology transfer, technology diffusion, and R&D should change over time in response to changing external environment. (2) Technology policies become effective only when three major components – policies to promote the demand side of technology, policies to promote the supply side of technology, and policies to provide effective links between demand and supply sides – are well balanced. Finally, it discusses implications for other developing countries.

Correspondence to: Linsu Kim, College of Business Administration, Korea University, Seoul 136-701, South Korea.

* The research underlying this article was supported by the World Bank. The authors are solely responsible for the views and interpretations expressed in this paper. In particular, these views and interpretations should not be attributed to the World Bank, to its affiliated institutions, or to any individual acting on behalf of these organizations.

Research Policy 21 (1992) 437–452
North-Holland

1. Introduction

Technological change is a major agent of economic growth. Development and adoption of new technology are the primary means by which scientific and technological advances are adapted not only to meet human and societal needs and expand new opportunities, but also to sustain and improve international competitiveness and economic growth. In industrialized countries, many studies have shown that more than 50 percent of long-term economic growth stems from technological changes which improve productivity or lead to new products, processes, or industries [8]. For this reason, a question that is often raised is how can science and technology also be used effectively for industrial development in developing countries. Industrial development is, in fact, a process of acquiring technological capabilities in the course of continuous technological change [25]. And science and technology policy (hereinafter technology policy) is a set of instruments the government uses in promoting and managing the process and direction of acquiring such capabilities. [1]

[1] See Kim [19] for a conceptual framework of how such capabilities may be acquired and Kim [17,18] for a detailed discussion of Korean experience.

Recognizing the importance of science and technology for industrialization, most developing countries have adopted a technology policy as a statement of intention. To provide the connecting link between the intention and its actualization, they have introduced various policy instruments as the vehicle to induce individuals and enterprises to make decisions following the lines established by the state. Important questions then, are: Do these instruments adequately cover all important aspects related to the promotion of science and technology? Are they effective as intended?

This paper is aimed at responding to such inquiries. Its central focus is to present an integrative framework, whereby technology policy instruments may be evaluated for their coverage and effectiveness. The paper, then, attempts to test its applicability as an analytical tool by using the framework to evaluate Korea's policy instruments. Given a phenomenal industrial growth in the past decades, Korea's technology policy may itself be an informative subject that may offer useful implications for policy makers in other developing countries.

2. Toward an integrative framework to analyze technology policy

Technology policy in this paper is defined broadly. It includes not only the direct instruments but also the indirect instruments that affect scientific and technological development. The former shapes the direction and pace of the supply side of technology development by strengthening technological capability. It is usually referred to as science and technology policy in the narrow sense of the term. The latter shapes the demand side of technology development by creating market needs for technological change and by providing various tax and financial incentives to lubricate linkages between supply and demand. This is usually referred to as industrial policy. An integrative framework will be developed below to analyze both direct and indirect instruments.

2.1. Market mechanism perspective

Kim [16] suggests that public policies to promote science and technology may be assessed by the market demand and supply for technology perspective. This perspective divides the categorization of science and technology policies into three major components: (i) policies designed to strengthen the supply side, increasing science and technology capabilities; (ii) policies designed to strengthen the demand side, creating market needs for technology; and (iii) policies designed to provide effective linkages between the demand and supply sides, attempting to induce and make innovation activities successful technically and commercially [21].

Unless there is a competitive market, where firms perceive innovation for products and processes as imperative to sustaining and raising market competitiveness so that they might survive and grow, there will be little investment in innovation activities, which are usually uncertain and risky. Also, capabilities established in public R& D institutes and universities will not be used efficiently in the absence of societal demands for such services. In this sense, science and technology policies should be an integral part of the overall industrial policies that shape market structure and industrial development.

However, even though the market calls for the introduction of new products and processes, if there exist no indigenous technological capabilities, countries cannot be expected to grow industrially. Local technological capabilities are necessary to insure that there is a supply response to the demand for new products, better quality and more efficient production.

Finally, despite the presence of both demand for innovation and supply of capabilities, few innovations can be expected to take place unless there is good management of the R&D system, effectively linking demand with supply. The linkage lubricates the innovation process. The absence of this linkage explains why in some industrialized countries there is little innovation despite a strong demand for innovation and an adequate supply of technical capabilities. The absence of this demand and supply of technology also explains why some policy instruments for linkage such as tax incentives for R&D efforts in developing countries are not effective in stimulating technological activities.

2.2. Technology flow perspective

This perspective is mainly concerned with three key elements in the flow of technology from abroad to a developing country [5,14]. The first is acquiring foreign technology. Technology transfer from abroad through such formal or direct mechanisms as multinational companies' direct foreign investment, the purchase of turnkey plants, patent and know-how licenses, and technical services may be important sources of technology that can facilitate the acquisition of technological capability in developing countries. Transfers of technological knowledge may also take place in informal modes such as imports of machinery (an extremely important mode of technology transfer) and valuable technical assistance provided by OEM (original equipment manufacturer) purchasers in an attempt to ensure that products meet their own stringent specifications [17]. In addition, Kim and Kim [20] report that sending nationals abroad for education, training, and work experience, to consult with technical journals, and to copy foreign products also play an important role in the informal transfer of technology that can be used by local firms in developing countries to acquire technological capability. In short, technology transfer from abroad by either formal or informal means is a major source of technological change in developing countries throughout the industrialization process.

Second, the effective diffusion of imported technology within an industry and across industries is another important element in upgrading technological capability of an economy. If a technology is transferred to a firm and its use is limited only to its original importer, it may give the firm monopoly power over other firms for a period of time; however, the economic effect of the technology may be considerably limited. If the economic benefits of technology transfer are to be maximized for a country, imported technology has to be diffused within an economy. Kim's [11] study of the electronics industry in Korea shows that the rapid diffusion of imported technology within the industry, through the mobility of experienced technical personnel, brought about the fast upgrading of technological capability by late comers, which in turn increased market competition, leading to increased investment for local technological efforts.

Third, local efforts to assimilate, adapt and improve imported technology and eventually to develop one's own technology are crucial to augmenting technology transfer and for expediting the acquisition of technological capability. Technology may be transferred to a firm from abroad or through local diffusion but cannot be assimilated effortlessly. Although knowledge may be transferred through various means, the ability to make effective use of it cannot. This ability can only be acquired through indigenous technological effort, leading to the acquisition of technological capability. These local efforts such as copying or reverse engineering of foreign products and processes, self-directed efforts aimed at improving and adapting previously acquired technology, and doing one's own research and development become increasingly important as industrialization progresses. These activities are necessary to strengthen international competitiveness in the face of increasing pressure from other developing countries.

These three elements are not necessarily sequential; rather, they may occur simultaneously. They are, however, important issues for policy makers to address explicitly in designing their technology related policy instruments.

2.3. *A dynamic perspective*

Many studies of industrialized economies have postulated the concept of a technological trajectory, delineating a multi-stage evolutionary path of technological development in response to a changing industrial environment [1,6,24,27]. Such a trajectory is important for understanding how product/process innovations and technology policy are inextricably linked as an industry and its technology evolve over time.

At the risk of oversimplification, the evolutionary path of a technology in industrialized economies can be categorized into three stages; emergence, consolidation, and maturity [7]. That is, industries evolve in these economies from the emergence stage (where product technology changes rapidly, innovating product performance drastically with high risk of technical/commercial failure) to the consolidation stage (where process technology changes rapidly, driving production costs swiftly down) and finally pass into the mature stage (where improvements in critical parameters become difficult to achieve, making industries increasingly inflexible and vulnerable in their competitive position).

In contrast, industries in developing economies initially begin at the mature stage of an industry and reverse their evolutionary direction preceding on to the consolidation stage [12,23]. For example, newly industrializing economies (NIEs) such as Korea, Taiwan, Hong Kong and Singapore began their industrialization in the 1960s, based largely on simple technologies at their mature stage and sustained their comparative advantage in labor-intensive mature industries by rapidly acquiring production capability. Well-trained but low-waged human resources, together with appropriate state policy and entrepreneurial dedication to accomplish their ambitious objectives, have worked together to achieve a rapid industrial progress in the 1960s and 1970s. Major industries represented include textiles, garments, and consumer electronics in the 1960s, and steel, petrochemicals, shipbuilding and automobiles in the 1970s. Technology and market for these industries had been proven elsewhere.

Economic environment for these NIEs has changed significantly in the 1980s for several reasons. First, in the wake of a rising trade imbalance, North America and Europe have moved toward protectionistic policies, that have made it increasingly difficult for NIEs to sustain export growth in industries that previously spearheaded their outward-looking strategy. Second, these countries have also put pressure on Korea and Taiwan to liberalize their domestic economies, forcing them to compete against multinationals not only in an international market but also within the domestic market (Hong Kong and Singapore have long opened their domestic market to multinationals). Third, these countries rapidly lost their competitiveness in low-wage-based labor-intensive industries, as their real wage rate has risen faster than productivity, making it difficult to compete against other developing countries with much lower wage rates. Fourth. industrially advanced countries (IACs), particularly Japan, are increasingly reluctant to transfer technology to NIEs, as the latter attempts to enter industries in the consolidation stage that have been dominated by these IACs. Fifth, NIEs were forced to change their copyright and patent laws to honor foreign intellectual property rights, preempting imitative reverse engineering of foreign products. As a result, many of the NIEs' industries appear to have entered the consolidation stage, in which they rely increasingly on technology to sustain their competitiveness. To cope with these environmental changes, the governments were forced to make a major shift in technology policy. In short, since a different stage in technological evolution calls for a different set of policy instruments, it is imperative to analyze technology policy from a dynamic perspective.

2.4. *An integrative framework*

The integrative framework consists of combining the aforementioned three frameworks into a

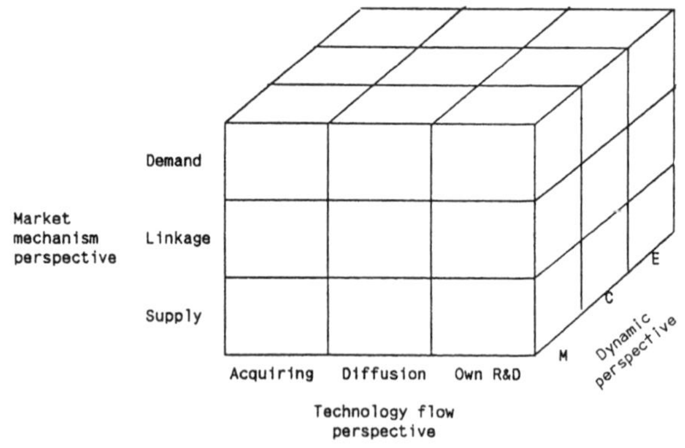

Fig. 1. An integrative framework. E, emergency stage; C, consolidation stage; M, mature stage.

matrix with three dimensions. The market mechanism and the technology flow approaches are combined into a matrix as illustrated in figure 1. The dynamic perspective is added to the other two as a third dimension to indicate time. The time dimension is important because its relative emphasis on acquiring, diffusing, and developing technology, as well as on the policies and mechanisms used to foster the demand, supply, and linkage aspects change as an economy progresses through different stages over time.

Such an integrative framework enables policy makers to identify the areas that have been neglected and to assess the efficacy of policy instruments presently deployed. The following Korean case shows that the framework sheds light on the effectiveness of intended as well as unintended policy instruments.

The experience of Korea will be used to demonstrate the usefulness of this framework. Few industries in Korea have yet entered the emergence stage and to simplify the exposition, only two stages – early (maturity) and recent (consolidation) – will be considered when analyzing technology policy for each of three key elements; demand, supply and linkages within the context of technology transfer, diffusion and its own R&D.

To cover each cell in the framework, however, forces repeated discussions of the same policy instruments under different cells, making it difficult to present the instruments in a logically readable manner. Nevertheless, the authors decided to highlight the framework at the risk of complicating the presentation of the Korean case, as they thought the original framework should be the main focus of this paper.

3. Korea's technology policy in the mature stage

Korea, as the case in point, has achieved a phenomenal economic growth in three decades through the rapid acquisition of technological capability. Thirty-six years of Japanese colonialization left some industrial base, mostly in the northern part. However, the political and economic vacuum and the chaos caused by the fall of Japanese colonial rule, the arbitrary division of the nation into North and South, and the ensuing civil war, all of which took place between 1945 and 1953, flattened Korea as a nation with a shattered past and a bleak future. As late as 1961, Korea suffered from almost all the difficulties facing most poor countries today. But beginning in 1962, the Korean economy grew at an average annual growth rate of 9 percent, raising the GNP per capita in current prices from $87 in 1962 to $5,569 in 1990. Exports increased from a mere $40 million in 1953 to $63 billion in 1990. The share of manufactured goods increased from 14.3 percent to 95.0 percent during the same period.

Before discussing specific policy instruments under the integrated framework, it is necessary to cover here three general but important points – administrative apparatus, human resource development, and big business – which are related so broadly to all three elements of technology flow that it is not appropriate to discuss them under specific sections below.

First, the Korean government established in 1967 the Ministry of Science and Technology (MOST) as a central agency to coordinate technology-related activities of other ministries. But its activities were largely ignored by action-oriented ministries that shaped industrial policies in promoting production and exports in labor-intensive mature industries. Consequently, MOST's long term science and technology policies were not integrated into national development plans. The National Council for Science and Technology was established in 1973 to bring about interministerial coordination, but met briefly only once upon inception and never functioned thereafter.

Second, a large part of Korea's success in acquiring, diffusing, improving and developing technology in the mature stage is due to its heavy investment in human resources. At the end of the Japanese colonial rule in 1945, only 2 percent of the Korean population over 14 years of age had completed secondary school and the illiteracy rate stood at 78 percent. Enrollments in the various levels of the formal education system have increased rapidly since then. For example, enrollment increased over five times in elementary school, 28.5 times for secondary school and almost 150 times for tertiary school, increasing drastically the percentage of school enrollment to the corresponding age group.

A study shows that compared to other developing countries, Korea was furthest in the direction of more education than would be expected, given GNP per capita from the regression line of 78 developing countries [9]. That is, with a per capita income of $90, Korea's educational achievement stood fairly close to the normal pattern of human resource development for a country with a mean per capita GNP of $200. Korea's level of human resource development with a per capita GNP of $107 was equivalent to that of countries with a GNP per capita of $380. As a result, Korea surpassed other newly industrializing countries by almost all indices of educational attainment [28]. Such a heavy investment in human resource development in early years resulted in a short-term unemployment problem of the educated but laid an important foundation for the subsequent economic development, which soon absorbed the surplus.

Third, the government intentionally created large firms, *chaebols*, as an instrument to bring about the economy of scale in mature technologies and in turn to develop "strategic industries" and to lead exports and economy. The government helped the capital formation as well as the subsequent diversification of these *chaebols*. The state sold Japanese colonial properties and state-owned enterprises to selected local entrepreneurs with a favorable long-term installment payment during the inflationary period, giving local entrepreneurs windfall capital gain. Owning all commercial banks, the government helped more significantly the formation of *chaebols* by allocating scarce foreign exchange and preferential financing at the official rates both of which were only one-half the real market rate [10]. The state also gave them large import-substitution projects, for which these entrepreneurs imported production technology on turnkey basis with foreign loans, which were guaranteed by the government. Then, foreign debt burdens due to currency devaluations were compensated with increased low interest loans, reducing market risks for these businesses [18].

Where Korea differs from other developing countries in promoting big businesses was in the discipline its state exercised over these *chaebols* by penalizing poor performers and rewarding only good ones. Good performers were rewarded with further licenses to expand. For those entering risky industries, the government rewarded entrants with other industrial licenses in more lucrative sectors, thus leading to further diversification. In contrast, the government refused to bail out relatively large-scale, badly managed, bankrupt firms in otherwise healthy industries, instead selected better managed *chaebols* to take them over [2].

3.1. Acquiring foreign technology

As postulated in the framework, foreign technology transfer may be facilitated or hindered through the presence or absence of policy instruments and their effectiveness in strengthening the demand, supply and their linkage.

First, the most important of all for creating *demand* for foreign technology in the mature stage may be one related to import substitution – domestic production of goods that were formerly imported, under the shelter of a high tariff wall and often under monopoly conditions – which created the need to import foreign technology. Since there was no local capability to establish and operate production systems, local entrepreneurs had to rely completely on foreign sources for production processes, product specifications, production know-how, technical personnel and component parts. Studies in the electronics [11], machinery [3], steel [13], computer [21], pharmaceutical [22] industries demonstrate that import substitution under protection was a most powerful instrument that has facilitated technology transfer from abroad, leading to the emergence of new industries and the introduction of more sophisticated products in the existing industries.

Korea's outward-looking policy appears to have been another important mechanism affecting the demand side of foreign technology transfer. While the Korean government highly protected the local market to foster infant industry growth, it promoted exports as something of a "life or death struggle" in order to bring about economic growth goals within a small domestic market. Thus, the government pushed and pulled firms with threats and promises [19]. Export promotion continually places pressure on firms to acquire foreign technology and to use it effectively in order to be able to compete in foreign markets. For example, lump-sum investments, for capacity beyond local market size, to bring about the economy of scale forced local businesses to acquire foreign technologies through both formal and informal mechanisms. They did so to maximize capacity utilization by improving international competitiveness. Consequently, the export-oriented industries accounted for the majority of licensing and capital goods imports in Korea. At the same time, they were also the prime beneficiaries of informal technology transfer, as foreign OEM buyers provided informal technical assistance in order to ensure Korea-made products met their technical specifications.

Second, despite demand for foreign technology, Korea's policy on the *supply* side of formal technology transfers such as direct foreign investment (DFI) and foreign licensing (FL) was quite restrictive in the early years of industrialization. When manufacturing accounted for only 12 percent of GDP, exports accounted for only 0.27 percent GNP, and primary goods accounted for 65.4 percent of total exports, technology was not a critical issue and the needed mature technologies could be acquired through mechanisms other than DFI or FL. Instead, Korean government promoted technology transfer through the procurement of turnkey plants and capital goods. Government policy was biased in favor of foreign capital goods as a way to strengthen international competitiveness of the capital goods user industries. Such a policy led to massive imports of foreign capital goods at the cost of retarding the development of the local capital goods industry [13]. Protection of the machinery industry was relatively low until the first half of 1971, giving capital goods users almost free access to foreign capital goods.

As a result, DFI had a minimal effect on Korean economy. For example, DFI's contribution to the growth of GNP in Korea in the 1972–1980 period amounted only to 1.3 percent, while its contribution to total and manufacturing value added was only 1.1 percent and 4.8 percent, respectively, in 1971, and 4.5 percent and 14.2 percent, respectively, in 1980 [4]. Most foreign licensing in early years were associated with technical assistance needed to train local engineers to run turnkey plants.

Third, various instrument also played an important role in providing *linkages* between the demand and supply of foreign technology transfer, lubricating the flow of foreign capital goods to Korea. For example, the slight overvaluation of the local currency, tariff exemptions on imported capital goods, and the financing of purchases by supplier's credits which carried low rates of interest relative to those on the domestic market all worked to increase the attractiveness of capital goods imports [26].

Table 1 shows that capital goods imports far surpassed other means of transfer in terms of value. Although the values of different modes of technology transfer are not strictly comparable since they measure different things, there are strong indications that compared to such NICs as Argentina, Brazil, India and Mexico, Korea has acquired more technology from IACs through the importation of capital goods than through any other means [28].

Contrary to policy intention, public research institutes played a limited role in bridging the needs of local industries with foreign technology. But in some cases, they played a consultative role for industries by identifying technology sources and undertaking RD&E (research, development and engineering), enabling local firms to strengthen their bargaining power against foreign technology suppliers.

Table 1
Foreign Technology transfer to Korea (US$ millions)

Source	62–66	67–71	72–76	77–81	82–86	87–89
Direct foreign investment	4.4	218.6	879.4	720.6	1,766.5	3,433.2
Foreign licensing	0.8	16.3	96.5	451.4	1,184.9	2,130.3
Technical consultancy	–	16.8	18.5	54.7	332.3	679.0
Capital goods imports	316	2,541	8,841	27,978	44,705	52,155

Source: Ministry of Finance for data on direct foreign investment and foreign licensing; Ministry of Science and Technology for technical consultancy data; and Korean Society for Advancement of Machinery industry for capital goods data.

3.2. Diffusion of technology

The Korean government did not have explicit policy instruments designed specifically for diffusion of technology. However, several unintended policy instruments had direct effects on demand, supply and linkage of technology diffusion.

First, import substitution also created *demand* for diffusion by inducing subsequent entries of new firms, which poached experienced personnel from existing firms, thereby bringing about the diffusion of imported foreign technology. A high mobility of experienced managerial and technical personnel has been a major problem for existing firms but at the same time it has been a major mechanism for technology diffusion for subsequent entrants. Kim [11] documented that two-thirds of the firms in the electronics industry during the 1960s and early 1970s acquired production and product design capability by poaching experienced technical personnel from the existing firms. A similar pattern is also evident in the 1970s in other mature industries such as in textile, machinery, shipbuilding, chemicals (see [13]).

The government's export-drive also made an important impact on the demand of technology diffusion even in early years. The mobility rate of experienced technical personnel was higher in outward-looking industries than inward-looking ones, as export markets grew faster in the 1960s and 1970s than the local market. Market incentives attached to the industrial standards system also affected the demand for the diffusion of imported foreign technology.

Second, on the *supply* side, specialized diffusion agents such as capital goods producers, consulting engineering firms, and public research institutes are the major mechanisms the government could use to promote the diffusion of technology within the economy. These agents were not, however, effective in diffusing technology in the early years.

The government's plan to develop the capital goods sector was first initiated in 1968 but was not seriously implemented until the mid-1970s. The development of local consulting engineering firms was promoted by the Engineering Service Promotion Law of 1973, which stipulated that all engineering projects, if possible, should be given to local firms as major contractors with foreign partners as minor participants. Such a scheme was aimed mainly at stimulating the emergence of local engineering firms and at providing local firms with opportunities to learn from experienced foreigners. But local infant engineering service firms were not capable of playing the role of diffusion agents in the early years of industrialization.

The government established in 1966 a public research institute as a diffusion agent. But Korean researchers, mostly from academic fields or from R&D centers in IACs, lacked the manufacturing know-how and the expertise to develop prototypes which were in the greatest demand during the early years; therefore, they could not compete with a foreign licensor in supplying detailed blueprints.

The most important diffusion agents the government unintentionally created were the state enterprises established in early years. Engineers, who accumulated modern production experience in state-owned fertilizer and machinery plants, spun off later to head engineering and production departments of private enterprises.

Third, the government established in 1962 a scientific and technological information center as a *linking* mechanism by disseminating technical information, but its use by industries was very limited in the early years, because mature products were easily imitated through reverse engineering without consulting technical literature.

3.3. Improving and developing technology

During the period of the mature stage, when relevant knowledge was readily available in a machine-embodied form and learning by doing was relatively easy, none of the policy instruments were effective in stimulating demand for or to strengthen supply of the country's own research and development (R&D).

First, there was a nearly total absence of *demand* for its own R&D to improve imported technologies and to develop its own technologies during the mature stage, a period when Korea enjoyed comparative advantage in labor-intensive light industries. The government also had no policy instruments to create demand to establish its own R&D. Action-oriented government agencies were, as mentioned earlier, completely preoccupied with achieving export and growth targets in these mature industries.

Second, in the absence of a demand for innovation activities, various efforts of the government (mostly by MOST) to *supply* technological capabilities in early years had questionable results. For example, the government aggressively recruited overseas-trained Korean scientists and engineers to establish a public R&D institute in 1966 as an integrated technical center to meet industry's technical needs. But there was no demand from industries for the kind of expertise the institute could offer, while the latter lacked the manufacturing know-how that would enable them to solve teething problems of the former in the early years.

The government took initiatives in indigenous R&D efforts, accounting for 70 percent of the nation's total R&D investment, but the total R&D investment accounted only for 0.39 percent of GNP in 1970. The private sector faltered in R&D investment despite the government's strong encouragements. This is evident from the fact that there was only one corporate R&D center in 1970. Universities, which accounted for 85 percent of high calibre scientists and engineers in Korea, did little research in the absence of demand for innovation.

Third, the government offered various tax incentives and preferential financing for R&D activities, but these *linking* mechanisms were largely ignored by industries due to the absence of a clearly felt need to invest in R&D given the relatively easy means of acquiring and assimilating foreign technologies then available from many sources.

In summary, when a major technological task was to undertake reverse engineering of mature foreign products through learning by doing, direct policy instruments (e.g. the creation of MOST and instruments to promote R&D investments) were ineffective. Rather, indirect instruments (e.g. expansion of formal education, import-substitution, export-promotion, imports of foreign capital goods, etc.) were more effective (1) in transferring foreign technologies informally and diffusing imported technologies throughout the economy and (2) in fostering infant industries to grow rapidly and to challenge international markets. That is, the continuous inflow of foreign technology, the formation of well-trained human resources with the absorptive capacity to undertake reverse engineering tasks, the entrepreneur's aggressive strategy to maximize learning under managerial independence from multinationals, the government's orchestral role in directing firms and selectively allocating resources to them to achieve ambitious growth objectives under the pressure of the competitive international market all have worked together to enable Korea to grow rapidly through the 1970s. In the absence of demand for technological change, direct instruments to strengthen the supply of technological capability or to provide linkages were ineffective.

4. Korea's technology policy in the consolidation stage

Korea's economic environment, as mentioned previously, has changed significantly in the 1980s. Recent protectionism in North America and Europe has made it difficult for Korea to sustain its export growth in mature industries. Increasing pressure to liberalize its domestic economy has forced Korea to compete against multinationals in both international and domestic markets. The rapid rise of real wage in Korea and concomitant entry of low-waged developing countries into the market place has resulted in the erosion of Korea's price competitiveness in mature industries. The increasing reluctance to transfer new technology to Korea and pressure to subscribe intellectual property rights to preempt imitative reverse engineering has forced Korea to secure alternatives to foreign sources for new technologies. To cope with these environmental changes, the government has to make a major shift in technology policy.

Again, before discussing specific policy instruments under the aforementioned integrative framework, changes in the three important aspects – administrative apparatus, human resource development, and big business – will be covered.

First, in light of the increasing importance of science and technology as an integrated national effort in the 1980s, the Korean government introduced a new administrative apparatus in 1982 – a quarterly Presidential Conference for Promotion of Science and Technology, similar to a pattern adopted in the 1970s to achieve export targets.

The incumbent administration scaled it down in 1990 to a President's Science and Technology Advisory Council. Although these two administrative apparatuses were designed to provide coordination between ministries involved in science and technology on a regular basis, day-to-day coordination has not been properly provided. This fact in turn has led to serious territorial conflicts between MOST and the Ministry of Trade and industry (MTI), the focal action-oriented ministry, which now regards science and technology as a major underlying factor in trade and industry.

Second, as the focal technological task has shifted from imitative reverse engineering to own innovation in the 1980s, the greatest focus in human resource development has been on higher education. Enrollment ratios in higher education increased over 70 percent in six years between 1980 and 1986. As a result, the number of scientists and engineers in Korea has more than doubled since 1980. Overseas training, which has been a characteristics of human resource development in Korea from the foreign aid period in the 1950s, also increased significantly: the ratio of third-level students training abroad to all third level students has been twice as high in Korea as in Argentina, Brazil, India and higher than in Mexico [28].

It should, however, be mentioned that the Korean government failed to develop in time an adequate stock of highly trained scientists and engineers who should be able to play a major role in the consolidation stage in competition against multinationals from IACs. A recent move by the government to enhance graduate programs resulted in across-the-board expansions in quantity, the environment for quality education and research has, however, deteriorated; the student–professor ratio has retrogressed from 22.6 in 1966 to 35.8 in 1985, making all universities primarily undergraduate teaching-oriented rather than graduate research-oriented. In short, Korea has failed to develop the research-oriented science and engineering manpower needed in the 1990s to sustain its competitiveness [19].

Third, the formation of *chaebols* appears to have helped bring about the economy of scale in the stable, mature industries, but in the face of *chaebols'* increasing economic power and its resultant monopolistic abuses (e.g. creating scarcities, price gouging, and ruining smaller competitors), the government adopted a policy of "economic democratization" to mitigate the monopolistic abuses of *chaebols*. The recently introduced Fair Trade Act along the lines of American antitrust legislation includes, among other things: the prohibition of unfair cartel practices and mutual investment among *chaebol*'s affiliate companies; sets a ceiling of credit flow going to predesignated *chaebols*; and regulates their vertical and horizontal integration. The government also began promoting small and medium-sized industries (SMI), particularly technology-based small firms, to remedy the imbalance between the large and small business sectors [15]. The government established "sanctuaries" for SMIs, designating 205 business territories where neither the large corporations nor their affiliates can intrude. Nevertheless, *chaebols* have grown even bigger and economic concentration increased further: the combined sales of the five largest *chaebols* as percent of GNP increased from 12.8 in 1975 to 35.0 in 1980 and to 52.4 in 1984.

4.1. Acquiring foreign technology

In order to sustain its international competitiveness in export-led industrialization even in the 1980s, the Korean government had taken several measures to progressively induce more sophisticated foreign technologies than those in the early years. Several policy changes in the 1980s affected the demand and supply side of foreign technology transfer and linkages in its process.

First, the government has played an important role by stimulating the *demand* side of technology transfer through procurement. For example, the government announcement that it would purchase 5,000 personal computers in 1982 and more in the subsequent years for public schools attracted many new entrants to the computer industry and induced existing firms to make aggressive investments for technology transfer, local R&D efforts and production, marking a major turning point for the industry. A more recent example is the government's move to computerize various nationwide systems [21], creating a market for various types of computer and software.

The recent revision of intellectual property laws that recognize both material and process patents preempts imitative product development which was possible under the previous process patent system. Such a policy shift also creates a significant amount of demand for transfer of foreign technology in the form of DFI and FL in the pharmaceutical and fine chemical industries.

The gradual liberalization of the domestic market under the pressure of the United States makes it imperative for Korean firms to enhance competitiveness by elevating their technological capabilities through foreign technology transfer as well as through intensifying their own R&D efforts.

More generally, the emphasis on exports continually places demand pressure on firms to acquire foreign technology and to use it effectively to compete in foreign markets. For example, lump-sum investment for capacity beyond local market size to bring about the economy of scale forces local businesses to acquire foreign technologies through both formal and informal mechanisms so that they might maximize capacity utilization by improving international competitiveness. Examples include large machinery makers, shipbuilders, and semiconductor producers.

Second, on the *supply* side of technology transfer, the Korean government reversed its policy on direct foreign investment and foreign licensing, substantially liberalizing guidelines in the beginning of 1980 as a way to induce more sophisticated foreign technologies through formal mechanisms. For DFI, the system of "positive list" (listing industries open to DFI) was switched to that of "negative list" (listing those that were prohibited), raising the number of 4-digit ISIC industries from 521 to 660. Terms and conditions have also been relaxed very much, allowing foreign suppliers greater ownership control. For FL, the policy was completely relaxed to allow foreign licensing for all industries and for all terms and conditions. As Korea becomes an important world competitor, it finds it increasingly difficult to obtain the advanced foreign technology it needs unless it allows foreign suppliers greater ownership control or favorable terms.

Third, on the *linkage* side, the government introduced various agencies to help industries to acquire foreign technology. (1) The government established a technology transfer center, which provides industries with information regarding alternative technologies available abroad and their suppliers and at the same time assists local firms to prepare contractual documents. (2) Three technical information centers collect and disseminate technical information, providing both transfer and diffusion effects. The subscription of such services by the private sector significantly rose in the 1980s. (3) Public R&D institutes often serve as a useful transfer agent. They undertake joint research with industries, enabling the latter to have enough technical knowledge to help them to identify foreign technology and its suppliers and to have bargaining power in negotiating technology transfer. Under the reversed policy environment that liberalized the flow of formal technology transfer and that created the demand for new technology, formal technology transfer such as DFI and FL increased drastically in the 1980s: Table 1 shows that the last three-year period (1987–89) accounted for over 48 percent of total DFI and almost 55 percent of total royalty payments. Capital goods imports have also risen drastically in order to improve the productivity of existing industries and to enter new industries.

4.2. Diffusion of technology

Despite the absence of explicit policy instruments designed to facilitate technology diffusion, several unintended instruments appear to affect the demand, supply, and linkage of technology diffusion.

First, on the *demand* side, industrial policy measures that enhance demands for innovation activities (mentioned under section 4.3 below) also affect the demand side of technology diffusion, since such measures strongly induce the interfirm mobility of experienced technical personnel.

Second, development of capital goods producers, consulting engineering firms, and public research institutes affect the *supply* of technology diffusion during the consolidation stage. For example, (1) measures to promote the development of capital goods industries (i.e. quantitative import restrictions, import licensing, domestic content requirements, etc.) induced local producers to develop technological capabilities so as to meet the increasing demand for capital goods. As a result of these measures and the priority placed on developing heavy industry in the early 1970s, capital goods production grew from $600 million in 1970 to over $67 billion in 1988, increasing the ratio of local production to local consumption from 48 percent to 57 percent. (2) The government's drive to promote construction service exports to the Middle East during the late 1970s enabled local construction firms to gain invaluable experience in engineering, leading to the emergence of many new engineering firms in Korea. (3) Public R&D institutes, though suffering from a poor linkage with industries, played an important supply role in technology diffusion, as these institutes generated many experienced researchers at a time when research was not a tradition in universities and industries. These researchers spun off later to emerging corporate R&D centers to play key roles.

Third, in the face of increasing demand for new technology in the 1980s, two mechanisms appear to have provided a useful demand–supply *linkage* in technology diffusion. (1) The most important mechanism affecting technology diffusion, particularly among small and medium-sized firms, has been technical extension services offered by six public agencies. Service ranges from assisting firms to improve product quality to training and helping firms to implement microelectronics-based factory automation. (2) Technical information centers collect, process, and disseminate scientific and technical information. They publish several monthly periodicals covering various areas of engineering on new technology development, including patent registration both outside and inside the country. A technical information center has an extensive on-line data base including CAS for chemical, chemical engineering and energy, INSPEC for electrical, electronics and computer areas, PATN for worldwide patent information, COMPENDEX for engineering in general, and NTIS for the U.S. government-owned technical information. In the face of increasing demand for innovation activities, the subscription of technical extension and information services by industries has significantly increased in the 1980s.

4.3. *Improving and developing technology*

In contrast to the mature stage, Korea's own R&D to improve imported technologies and to develop its own ideas became critical in the consolidation stage, given the drastic change in the economic environment. The Korean government introduced various policy instruments to create the demand for and to strengthen the supply of own technological activities.

First, since *demand* creation is the most important aspect of stimulating R&D activities, the government used various means. Government policies that enhance market competition play a most important role in creating demand for improving and developing technology. For example, liberalization on imports and foreign investment alerts local firms to strengthen their competitiveness against multinationals by investing heavily in R&D. Also, the protection of intellectual property rights that preempt imitative reverse engineering makes it imperative for local firms to acquire foreign technology or to invest heavily in R&D in order to sustain their market share. The recent anti-trust move by the government and its restriction on the concentration of economic power by large firms also enhance market competition, calling for innovation activities.

The government's new scheme to promote the import-substitution of major import items is another instrument that creates demands for R&D. That is, the government designates specific target machinery, parts, and raw materials that should be localized for import-substitution. It then offers tax incentives, preferential financing, loan guarantees, and R&D subsidies in the form of joint government–corporate research to those who develop the designated items. Through 1987, the government has designated 1,555 such items.

Public procurement policies also create needs for technology development efforts by the private sector. Prior announcement of a medium-range (three year) procurement plan by public agencies was first introduced in 1984. The number of public agencies adopting the scheme increased from three in 1984 to eight in 1986, while the number of procurement items under this scheme increased from 409 to 1,174 and the total value from $206 million to $324 million during the same period. In 1986, the government adopted a new method in selecting successful bidders, whereby a quality index was added to price in the selection criteria. Such a method was designed to minimize the life cycle costs of procurement and to promote local technological efforts in improving product quality.

The protection of locally developed new technologies from entry by direct foreign investment or foreign licensing, and from local imitation for a period from one to four years also encouraged R&D efforts by the private sector. The government also offered preferential financing and preferential treatment in procurement. During the first seven years, only 11 cases were designated by the government, but 33 were designated in the 1985–86 period. This policy has, however, been phased out in order to create a consistent antitrust policy environment.

Second, public roles in strengthening the *supply* side of improving and developing technology vary. The government established several specialized research institutes as spin-offs from the initial one to develop in-depth capabilities in an area of high industrial priority in shipbuilding, marine resources, electronics, telecommunications, energy, machinery, chemicals, and standardization. These research centers have made a series of important contributions to industrial development by pioneering new products and processes and adapting and improving imported foreign technologies, as the needs of the private sector became more sophisticated to require assistance from expertise available from these centers.

The Korean government has also promoted R&D activities. Although the Korean economy has recorded one of the highest GNP growth rates, R&D expenditure rose faster than GNP, increasing its share of GNP from 0.32 percent in 1971 to 1.90 percent in 1988. Public R&D investment increased significantly from a mere 7.29 billion Won to 416.2 billion Won during the same period. But the public role in R&D activities declined noticeably relative to that of the private sector (see Table 2).

The most impressive fact is that there was such a massive increase in private R&D efforts. In the face of increasing market competition in both domestic and international markets and with various government incentives, private R&D investment exploded from a mere 3.38 billion Won in 1971 to 1.9 trillion Won in 1988, raising its share in total R&D expenditures from 32 percent to 82 percent over the period. R&D expenditures in the manufacturing sector, as a percentage of total sales, also increased from 0.36 percent in 1976 to 1.61 percent in 1988. The number of corporate R&D centers increased from one in 1971 to 122 in 1983 and to 749 by 1989.

Third, the main instruments to lubricate innovation processes that *link* demand for innovation and the supply of capabilities include tax incentives, preferential financing, direct R&D grants, and venture capital creation. Tax incentives include: reduced tariffs on import of R&D equipment; the deduction of annual non-capital R&D expenditures and human resource development costs from taxable income; accelerated depreciation on industrial R&D facilities; and a tax credit for investment in R&D facilities and for the

Table 2
Major R&D Indicators in Korea (Unit: billion Won)

	1971	1976	1981	1986	1988
R&D expenditure	10.67	60.90	293.13	1,523.28	2,347.4
Funds from government	7.29	39.18	121.73	289.44	416.2
Funds from private sources	3.38	21.72	171.40	1,233.84	1,931.2
Government vs. private	68:32	64:36	42:58	19:81	18:82
Manufacturing R&D	na	16.70	111.77	935.10	1,633.3
Manufacturing R&D/sale	na	0.36	0.67	1.36	1.61
GNP	3,376	13,881	45,126	83,833	123,579
R&D Exp/GNP	0.32	0.44	0.65	1.82	1.90
Number of researchers	2,477	11,661	20,718	47,042	56,545
R&D exp/researcher (thousand Won)	4,306	5,223	14,149	32,381	41,514
Researcher/per 1,000 population	0.08	0.33	0.54	1.33	1.35

Source: Ministry of Science and Technology, *Science and Technology Annual, 1987*, MOST, 1987 and Korean industrial Research Institute *Science and Technology Statistics* 1990.

commercialization of locally-generated R&D results. The incentives also include a tax reduction scheme, called the Technology Development Reserve Fund, whereby an enterprise can set aside up to 20 percent (30 percent for high-tech industries) of profits before tax in any one year to be used for its R&D work in the following four years. it was only in the 1980s that these instruments became effective in the wake of increasing market demand for technological innovation and the significant improvement of local technological capability.

Preferential financing for R&D activities has come from five major sources: a technology development fund earmarked within the National Investment Fund, Industrial Development Fund, the Korea Development Bank Technology Development Fund, industrial technology promotion funds earmarked specifically for automation and new material development, and the Small and Medium Industry Promotion Fund. In addition, the government enacted the New Technology Commercialization Financing Promotion Act in 1986, by which it promotes and invests in the formation of financial institutions specifically catering to the needs of new technology-based firms. As a result, preferential financing became the most important mechanism funding corporate R&D, accounting for 64 percent of the total R&D expenditure in manufacturing in 1987 [18].

The government also played an important role in facilitating local technological efforts by establishing the venture capital industry. In the absence of private interest in venture capital, the government enacted in 1981 a special law to establish a venture capital firm. There are now three public venture capital firms. Two majority government-owned banks also offer a limited venture capital service. Also, the government has taken a further step with the enactment of the Small and Medium Enterprise Formation Promotion Act in 1986, whereby 12 venture business firms, invested jointly by the private sector and the government, were established.

In summary, to cope with drastic changes in economic environment, the Korean government made a major policy shift in the 1980s. Various industrial policies such as import liberalization and the subscription of intellectual property rights, continued emphasis on exports, etc. were introduced in the 1980s to further enhance the demand side of technology transfer, diffusion, and R&D. The liberalization of DFI and FL, the promotion of the capital goods industry and consulting engineering, expansion of public R&D institutes, and drastic increases in R&D were new instruments to strengthen the supply side of technology development. Linking mechanisms such as tax incentives and preferential financing became catalytic in technology development in the dynamic presence of both demand for and supply of technological capability. However, territorial conflicts among government ministries in designing a new national system of industrial innovation, deficiencies in the higher educational system, and ever increasing dominance of *chaebols* in the Korean economy are some of issues the government has to address.

5. Conclusions and implications for other countries

The integrated framework to analyze technology policy and Korea's experience as presented above has several implications for other developing countries. First, technology policies become effective only when three major components – policies to promote the demand side of technology, policies to promote the supply side of technology, and policies to provide effective links between demand and supply sides – are well balanced. For example, Korea's technology policies to promote technological capability and various incentives for technological development efforts were not effective in the 1960s and 1970s in the absence of a strong market demand for technology. It was only in the 1980s, when technology was regarded as one of the most important underlying variables in market competition that the aforementioned policies and incentives became effective. This suggests that supply-side oriented technology policies should be an integral part of demand-side oriented industrial policies.

Second, policy instruments are the vehicle to induce individuals and enterprises to make decisions along the lines established by the government. But they are not necessarily effective as intended, with a significant gap between intended and realized effects. Policy makers should introduce objective means to evaluate the efficacy of intended instruments. Such an evaluation should occur regularly, as changes in environment affect their effectiveness.

Third, Korea's experience on technology transfer shows that it had very restrictive policies toward formal mechanism of technology transfer (i.e. DFI and FL) but relied more on informal mechanisms (i.e. through the importation of capital goods and reverse engineering) in early years. This was not only the decision of the private firms but also that of the government, as its approval was required for each DFI, FL, and turnkey plant agreement. Korea's approach had been effective in terms of the cost of obtaining foreign technology as well as maintaining its "independence" from the dominance of multinationals. Such a policy had both positive and negative effects. On the positive side was the preclusion of constraints often imposed by foreign multinationals on local firms' efforts to develop their own capability. On the negative side was the lack of continued access to new technologies that may have been possible through direct equity links with foreign firms. However, given the training and entrepreneurship of Koreans, this policy appears to have been appropriate in the mature stage. In short, such an approach is only possible when a country has highly trained human resources and entrepreneurship that enable it to assimilate and adapt foreign technologies embodied in physical items or technical information available in literature.

Fourth, it was Korea's heavy investments in human resource development that made it possible to acquire technological capability rapidly and, in turn, to achieve rapid industrialization during the mature stage. Korea invested heavily in human resources development even well before launching its industrialization program. The rapid expansion of education resulted in serious unemployment problems for the educated in the 1950s and 1960s, but the formation of educated human resources, despite the low per capita income, laid an important foundation for the subsequent industrial development by the rapid adaptation of imported technology and reverse engineering of foreign products. The rapid development of the Korean economy soon absorbed the surplus. On the contrary, Korea's neglect in investing in research-oriented graduate education is a serious bottleneck in sustaining its economic growth. Technological capability is embodied in human resources and has two elements: prior knowledge and intensity of efforts. Research-oriented graduate education will produce highly trained scientists and engineers with relevant prior knowledge in the technological frontier. In short, other policies related to technology promotion may be conceived and implemented in a short time span, but manpower development can only be built through long-term planning and investment. In this sense, the government should start investing for human resource development early enough for subsequent industrialization. The intensity of efforts is not covered in this paper, as it cannot easily be manipulated by policy instruments. It is, however, a critically important sociocultural factor in industrialization [19].

Fifth, even during the labor-intensive period of the 1960s, the government acted on its vision of an industrialized Korea by establishing a science and technology infrastructure at a time when demand for such an infrastructure was little in evidence. These programs had their own problems, but produced a large number of experienced, high caliber researchers, who have later been absorbed by other public and private R&D centers established in the late 1970s and 1980s. In short, Korea's experience in public R&D activities shows that the major role of public R&D institutes in the early days of industrialization should be that of generating experienced researchers and helping the private sector gain foreign technologies. Then, the role evolves to adaptive development and R&D in response to the economy's progressive advance to technology-intensive areas.

Sixth, the strong export-orientation of the Korean economy has facilitated the rapid acquisition of technological capability at least in two ways. (1) As producers entered the international market, the keen international competition forced them to invest more in technological efforts. (2) Informal technical assistance offered by foreign OEM buyers to ensure that Korea-made products met their technical specifications provided invaluable help to Korean firms in acquiring the necessary technological capability. Thus, there can be important positive externalities from a strong participation in international trade. A hypothesis is that inwardly-oriented developing countries may not be able to acquire technological capability as fast as outwardly-oriented countries, since the former may be under less competitive pressure.

Finally, can Korea's experience be transferrable to other developing countries? Korea has its own peculiarities; the availability of well-trained and disciplined human resources and entrepreneurship, sociocultural characteristics of its people, policies to promote exports and to restrict foreign investment, its vulnerability due to lack of natural resources, and characteristic modes in interactions between the state and enterprises. Thus, Korea's experience cannot easily be transferred to other developing countries but may serve as a useful lesson for them.

References

[1] W.J. Abernathy and J.M. Utterback, Patterns of Industrial Innovation, *Technology Review* (June/July 1978) 41–48.

[2] A.H. Amsden, *Asia's Next Giant: South Korea and Late Industrialization* (Oxford University Press, New York, 1989).

[3] A.H. Amsden and L. Kim, Technological perspective on the General Machinery Industry in the Republic of Korea, in: Martin Fransman (editor), *Machinery and Economic Development* (MacMillan Press, London, 1986) pp. 93–123.

[4] D.S. Cha, *Weja Doipeo Hyogwa Boonsuk* (The Effects of Direct Foreign investment) (KIET Press, Seoul, 1983).

[5] C.J. Dahlman, Technology Policy and Strategy in Developing Economies, Industrial Development Department, The World Bank, a mimeograph (1989).

[6] G. Dosi, Technological Paradigms and Technological Trajectories, *Research Policy* 11 (1982) 147–162.

[7] H. Ergas, Does Technology Policy matter? In: Bruce R. Guile and Harvey Brooks (editors), *Technology and Global Industry: Companies and Nations in the World Economy* (National Academy Press, Washington, DC, 1987).

[8] M. Goldsmith (editor), *Technological innovation and the Economy* (Wiley Interscience, London, 1970).

[9] F.H. Harbison and C.R. Myers, *Education, Manpower and Economic Growth* (McGraw-Hill, New York, 1964).

[10] K.H. Jung, Business–Government Relations in Korea, in: Kae H. Chung and Hak Chong Lee (editors), *Korean Managerial Dynamics* (Praeger, New York, 1989) pp. 11–26.

[11] L. Kim, Stages of Development of industrial Technology in a LDC: A model, *Research Policy* 9 (1980) 254–277.

[12] L. Kim, Technological Change and its Implications for Less Developed Countries, in: Dong-Ki Kim (editor), *Organizing for Higher Productivity* (Asia Productivity Organization, Tokyo, 1986) 219–232.

[13] L. Kim, *Imitation and Apprentice: How Korea Acquired Technological Capability Fast* (College of Business Administration, Korea University, 1987).

[14] L. Kim, *Korea's Technology Policy for Industrialization*, The Industrial Development Department, The World Bank, a mimeograph (1988).

[15] L. Kim, Entrepreneurship and Innovation in a Rapidly Developing Country, *Journal of Development Planning* 18 (1988) 183–194.

[16] L. Kim, Science and Technology Policies for Industrialization in Korea, in: Jangwon Seo (editor), *Strategies for Industrial Development: Concept and Policy Issues* (APDC, Kuala Lumpur, 1989) 255–272.

[17] L. Kim, Pros and Cons of International Technology Transfer: A Developing Country View, in: Tamir Agmon and Mary Ann von Glinow (editors), *Technology Transfer in International Business* (Oxford University Press, New York, 1991) 223–239.

[18] L. Kim, National System of Industrial Innovation: Dynamics of Capability Building in Korea, Working Paper WP 91–1 (Korea University, College of Business Administration); and also forthcoming in Richard Nelson and Nathan Rosenberg (editors), (Oxford University Press, New York, 1992).

[19] L. Kim, Absorptive Capability and Industrial Growth: A Conceptual Framework and Korea's Experience, a paper presented at the 20th Anniversary Symposium on Economic Growth and Social Capability in Seoul, Korea (1–3 July 1991).

[20] L. Kim and Y. Kim, Innovation in a Newly Industrializing Country: A Multiple Discriminant Analysis, *Management Science* 31 (1985) 312–322.

[21] L. Kim, J. Lee, and J. Lee, Korea's Entry into the Computer Industry and its Acquisition of Technological Capability, *Technovation* (September 1987) 277–293.

[22] Y. Kim, L. Kim, and J. Lee, Innovation Strategy of Local Pharmaceutical Firms in Korea: A Multivariate Analysis, *Technological Analysis and Strategic Management* 1 (1989) 29–49.

[23] J. Lee, Z.T. Bae, and D.K. Choi, Technology Development Processes: A Model for a Developing Country with a Global Perspective, *R&D Management* 18 (1988) 235–250.

[24] R.R. Nelson and S.G. Winter, *An Evolutionary Theory of Economic Growth* (Harvard University Press, Cambridge, MA, 1982).

[25] H. Pack and L.E. Westphal, Industrial Strategy and Technological Change, *Journal of Development Economics* 22 (1986) 87–128.

[26] Y.H. Rhee and L.E. Westphal, A Micro, Econometric Investigation of Choice of Technology, *Journal of Development Economics* 4 (1977) 205–237.

[27] W.H.C. Simmonds, Towards an Analytical Industry Classification, *Technological Forecasting and Social Chance* 4 (1973) 375–385.

[28] L.E. Westphal, L. Kim, and C.J. Dahlman, Reflections on the Republic of Korea's Acquisition of Technological Capability, in: Nathan Rosenberg and Claudio Frischtak (editors), *International Technology Transfer: Concepts, Measures, and Comparisons* (Praeger, New York, 1985) pp. 167–221.

[10]

National System of Industrial Innovation: Dynamics of Capability Building in Korea

LINSU KIM

South Korea (hereafter Korea) has emerged as one of the fastest growing economy and has been transformed from a subsistent agricultural country into a newly industrial one during the past quarter century. As late as 1961, Korea suffered from almost all difficulties facing most poor countries today. But beginning in 1962, the Korean economy grew at an average annual growth rate of almost 9%, raising GNP per capita in current prices from $87 in 1962 to $4936 in 1989. Exports increased from a mere $40 million in 1953 to $62 billion in 1989. The share of manufactured goods in exports increased from 14.3 to 95.0% during the same period. In the mid-1960s, Korea began exporting textile, apparel, plywood, and other labor-intensive products. Ten years later, ships, steel, and construction services from Korea challenged established suppliers from the advanced countries. By the mid-1980s, computers, semiconductor memory chips, video cassette recorders, and automobiles were added to the list of Korea's major export items. By the end of the 1980s, Korea's R&D investment as a fraction of GNP and the number of scientists and engineers as a fraction of the work force were approaching the levels of some of the highly industrialized countries of Europe. What made it possible for Korea to achieve such a phenomenal growth in industrialization? Will Korea be able to sustain the growth in the future?

As a way to answer these questions, this chapter presents Korea's national system related to industrial innovation, which influences to a great extent the direction and pace of technological capability development that in turn lead to industrialization. The main focus of this chapter will be concerned with describing the industrial innovation system as of this new era in Korea's industrial history. It is, however, also important to examine how the national system in the past has worked for Korea to acquire technological capabilities that brought the country where it is today.

HOW KOREA GOT WHERE IT IS TODAY

Thirty-six years of Japanese colonialization left some industrial base, mostly in the northern part. But political and economic vacuums and chaos caused by the fall of Japanese colonial rule, arbitrary division of the nation into North and South, and ensuing civil war, which all took place between 1945 and 1953, flattened Korea as a nation with little left of its past and facing a bleak future.

Particularly, the Korean War destroyed the majority of industrial and infrastructure facilities Korea inherited from Japanese colonialization. When the Armistice was signed in 1953, the net commodity product decreased by 26% compared to the figure 10 years before, decreasing net commodity product per capita by 44% (Kim and Roemer, 1979). Ironically, the Korean War also served positively to a certain extent for the subsequent economic development, having completely transformed a traditional rigid society into a highly mobile one by forcing geographical mobility and contributed to the rapid formation of basic skills among the male labor force, most of who served compulsory military service.

Given virtually no domestic savings, about 70% of all reconstruction projects had to be financed by foreign aid. American economic assistance averaged about 10% of Korean GNP between 1953 and 1960, enabling the country to recover back to the prewar period by the time the aid was completely phased out and Korea launched its economic development program in the early 1960s.

Korea's rapid economic progress may be attributed to many economic, social, and technical factors. The most important of all may be technological capabilities, which is a combined outcome of various economic, social, and technical inputs. It is the ability to make effective use of technical knowledge in production, investment, and innovation. Technological capability enables one to assimilate, use, adapt, change, or create technology and to develop new products and processes in response to a changing economic environment. The process of acquiring technological capability is so diverse and complex that it defies a simple analysis. But this section attempts to identify both macro/general and micro factors that influenced the acquisition of technological capability in Korea.

Macro/General Factors

Macro/general factors—the process of human resource formation, the inflow of foreign technologies, the government's industrial, trade, and science/technology policies that shaped industrial structure and the direction of growth, and so on—set the stage for the rapid acquisition of technological capability.

First, human resource development may be a most basic and crucial determinant, as technological capability is embodied in people. Korea's greatest resource is its human resources. Modern education was first introduced to Korea by American missionaries and later expanded by the Japanese colonial government. But at the end of the Japanese colonial rule, only 2% of the Korean population over 14 years of age had completed secondary school, and the illiteracy rate stood at 78%.

The importance of education in the Korean government's modernization strategy can be seen in the growth of government investment in education: the share of education in the total government budget, for instance, rose from 2.5% in 1951 to over 22% in the 1980s. Government expenditures, however, accounted for only one-third of the total expenditures in education, the remainder being borne by the private sector and parents, reflecting the high commitment for education held by the Korean society. This commitment was the strongest in Korea among eight industrialized (Denmark, Germany, Italy, Japan, Sweden, Switzerland, the United Kingdom, and the United

States) and two semiindustrialized (Singapore and Korea) countries Porter (1990) studied.

As a result, enrollment in the various levels of the formal education system has increased rapidly since 1953. In the case of elementary school, enrollment has increased by over five times. Even faster growth is seen in the secondary and tertiary education levels. Secondary school enrollment increased by more than 28.5 times between 1945 and 1980, while enrollment in colleges and universities increased by almost 150 times during the same period. Table 11.1 shows that school enrollment as a percentage of the corresponding age group rose to over 100% in 1970 for the elementary school level, to over 95% in the early 1980s for the middle school level, to over 70% for the high school level, and over 25% for the tertiary level by 1980.

Several other developing countries also attained as rapid growth in elementary education as did Korea. But what was unique in Korea was the well-balanced expansion in all levels of education early enough to support its economic development. Harbison and Myers (1964) show that given GNP per capita from a regression line of 73 developing countries, Korea's level of educational achievement was furthest of that predicted by its level of economic development. That is, with a per capita income of $90, Korea's educational achievement stood fairly close to the normal pattern of human resource development for a country with a mean per capita GNP of $200. Korea's level of human resource development, when its per capita GNP reached $107, was equivalent to that of countries with a GNP per capita of $380.

Overseas training and observation were also characteristic of human resource development in Korea. Overseas training traces back to the foreign aid period in the 1950s, when a high proportion of senior personnel in government, business, and academia were exposed to foreign training, mainly in the U. S., under economic assistance

Table 11.1. Indicators of Human Resource Development in Korea

	1953	1960	1970	1980	1987
Literacy rate (%)	22.0	72.1	89.4	[a]	[a]
Enrollment as percentage of corresponding age group					
Elementary school (ages 6–11)	59.6	86.2	102.8	101.0	100.2
Middle School (ages 12–14)	21.1	33.3	53.3	94.6	98.8
High School (ages 15–17)	12.4	19.9	29.3	68.5	82.8
College/university	3.1	6.4	9.3	14.9	25.5
Graduates of vocational training centers (1000)	NA	NA	31.6	104.5	50.9
Number of graduates from tertiary schools (per 10,000 population)	NA	10[b]	11	27	54
Number of scientists and engineers[c]	4,157	16,436	65,687	174,832	361,330

[a]Illiteracy rate after the mid-1970s was so insignificant that the government ceased to collect data on it.

[b]For 1965.

[c]Cumulative number of science and engineering graduates since 1945.

Source: Ministry of Education and McGinn et al. (1980).

programs (Mason et al., 1980). The tradition of overseas training continues to this day. The ratio of third-level students training abroad to all third-level students has been twice as great in Korea as in Argentina, Brazil, and India and higher than in Mexico (Westphal et al., 1985). As a result, Korea surpassed the aforementioned newly industrializing countries by almost all indices of educational attainment. The rapid expansion of education resulted in serious unemployment problems for the educated in the 1950s and 1960s, but the formation of educated human resources, despite low per capita income, laid an important foundation for the subsequent adaptation of imported technology and development of indigenous technology. The rapid expansion of the Korean economy soon absorbed the surplus of educated workers.

Second, Korean's hard working habits in long hours may be another factor accounting for the fast acquisition of technological capability. Korea's land area is scarcely larger than New Jersey or Hungary, crossed by mountains with a relatively small portion of the land arable for more than 40 million people. In terms of population density, Korea trails only Bangladesh and Taiwan. The cramped conditions and severely cold winters appear to have forced Koreans to work hard and long whenever possible in order to survive in an unfavorable environment. The average working hours per week in manufacturing in 1985 was 53.8 hours in Korea compared to 33.1 to 42.9 in OECD countries including Japan, 44 to 48.0 hours in other Asian NICs, and 46.0 hours in Mexico (Kim, 1988). Besides, the older generation is motivated by the memory of deprivation and hard times under Japanese occupation and during the Korean War. A sense of *Geug-il* or "beat Japan" to settle old scores and national economic competition with North Korea are also major forces motivating Koreans (Porter, 1990).

Third, lacking technological capability at the outset of its economic development, Korea had to rely on foreign technology imports. However, Korea's policies concerning direct foreign investment (DFI) and foreign licensing (FL) were quite restrictive in the early years of industrialization, when technology was not a critical element and the mature technologies needed could be easily acquired through mechanisms other than DFI or FL (e.g., reverse engineering). Table 11.2 presents data on technology transfer through DFI and FL to Korea.

Consequently, the size of DFI and its proportion to total external borrowing were significantly lower in Korea than in other NICs. For example, Korea's stock of DFI in 1983 was only 7% of the size of that in Brazil, 23% of that in Singapore, and less than a half the size of that in Taiwan and Hong Kong. The proportion of DFI to total external borrowing was only 6.1% in Korea compared with 91.9% in Singapore, 45% in Taiwan, and 21.8% in Brazil (KEB, 1987). The comparative figure reflects Korea's explicit policy of promoting its "independence" from multinationals in management control.

In contrast, Korea promoted technology transfer through the procurement of turnkey plants in the early years. For example, the chemical, cement, steel, and paper industries, established in the 1960s and early 1970s, all resorted to the turnkey mode for their initial setup. But Korean firms assimilated imported technologies so rapidly that they managed to undertake subsequent expansions and upgrading with little assistance from foreigners. The rapid growth of the Korean economy has called for commensurate growth in investment for production facilities. Government policy had, however, been biased in favor of the importation of foreign capital goods and against

Table 11.2. Foreign Technology Transfer to Korea (US$ Million)

Source	62–66	67–71	72–76	77–81	82–86	Total
1. Direct foreign investment (DFI)						
Japan	8.3	89.7	627.1	300.9	875.2	1,901.2
United States	25.0	95.3	135.0	235.7	581.6	1,072.6
All others	12.1	33.6	117.3	184.0	309.7	658.7
Total	45.4	218.6	879.4	720.6	1,766.5	3,632.5
2. Foreign licensing (FL)						
Japan	—	5.0	58.7	139.8	323.7	527.2
United States	0.6	7.8	21.3	159.2	602.7	791.6
All others	0.2	3.5	16.6	152.4	258.5	431.1
Total	0.8	16.3	96.5	451.4	1,184.9	1,749.9
3. Technical consultancy						
Japan	—	12.1	7.7	20.8	89.2	129.8
United States	—	3.1	6.0	16.7	159.1	184.9
All others	—	1.6	4.8	17.2	84.0	107.6
Total	—	16.8	18.5	54.7	332.3	422.3
4. Capital goods imports						
Japan	148	1,292	4,423	14,269	20,986	41,118
United States	75	472	1,973	6,219	12,394	21,133
All others	93	777	2,445	7,490	53,338	64,143
Total	316	2,541	8,841	27,978	86,718	126,394

Source: Ministry of Finance on DFI and FL; Ministry of Science and Technology for technical consultancy data; and Korean Society for Advancement of Machinery Industry for capital goods data.

the development of a rudimentary local capital goods industry as a way to strengthen international competitiveness of the capital goods user industries. Such a policy led to massive imports of foreign capital goods, which became a major source of learning through reverse engineering by Korean firms (Kim and Kim, 1985). Of the four categories of technology transfer listed in Table 11.2, capital goods imports far surpass other means of technology transfer in terms of value. Capital goods imports were worth 34 times the value of DFI and 72 times the value of FL. The total value of capital goods imports is 21 times that of the other two categories combined, suggesting that Korea may have acquired more technology from advanced countries through the importation of capital goods than through any other means. Among NICs, the proportion of capital goods imports to total technology transfer was highest in Korea (Westphal et al., 1985).

In short, Korea restricted DFI but promoted instead technology transfer through other means such as capital goods imports. Capital was acquired in the form of foreign loans. Such a policy, designed to maintain Korea's independence from foreign multinationals, has been relatively effective in acquiring technological capability in the mature industries, as Korea's well-trained human resources and their entrepreneurship enabled the country to learn quickly from foreign capital goods.

Fourth, continued relations with the United States in the national defense has served an important role in enabling Korean firms to accumulate experience and capability. The U. S. military procurement program afforded many producers in a number of sectors with occasions for assisted learning by doing to meet exacting product specifications. Construction contractors, plywood producers, and tire makers, among others, built up their initial capability through the U. S. military contracts and later became major exporters (Westphal et al., 1985). For example, through U. S. military contracts, local construction firms learned, under the direction of the U. S. Engineering Corps, specific details of stringent U. S. specifications, the preparation of bidding documents, and management of construction projects, enabling local firms to incrementally accumulate increasingly more sophisticated construction capabilities. By the early 1970s, Korean construction firms became important contractors for the U. S. military bases in Guam and Vietnam. Capabilities accumulated in these foreign countries led Koreans to launch an aggressive expansion to the Middle East in the wake of oil crises. The U. S. military presence also reduced the defense expenditure burden of the Korean economy, enabling the country to divert more resources for industrial development.

Fifth, despite a strong backup of the U. S. military presence in Korea, Korea still had to maintain one of the largest military forces in the Free World by allocating over 36% of its annual budget to national defense. Such a national security concern, particularly during and after the Carter Administration, which expressed a possibility of the U. S. military pullout from Korea, had some impact on the development of particular industries and their technological capabilities. For example, one of the major factors the Korean government has considered in pushing hard heavy machinery industries in the late 1970s appears to be its concern to acquire capability for defense production. Although such an ambitious investment in heavy machinery industries in the absence of adequate local capability has resulted at the outset in extremely low capacity utilization and resource allocation problems, the investment over local market and local capability has put enormous pressure on both the government and private entrepreneurs to upgrade technological capability for survival. As a result, Korea became an important contender in exporting large scale plants in the Third World. The major impact of national security concern in technological capability building may be the significant improvement in precision in the machinery industry. The procurement demand of the Korean military after the Korean War also helped local consumer electronics producers to diversify into industrial electronics.

Sixth, to overcome a disadvantage of a small-sized domestic market and to take advantage of the stable nature of mature technologies on which industrialization strategy was based, the Korean government designated so-called "strategic" industries for import substitution and export promotion. Plywood, textiles, consumer electronics, and automobiles in the 1960s and steel, shipbuilding, construction services, and machinery in the 1970s are examples. These industries, which were created in violation of its static comparative advantage, had to suffer from high costs in addition to "infant industry" growing pains. To enable the industries to overcome the problems, the government used various policy mechanisms; assistance to mitigate risk and to secure stability, market protection from foreign competition, export promotion, and tax incentives and preferential financing for products and producers designated by the government. The average effective rate of protection was atypically high for the "strategic" industries. In some industries, protection quickly became redundant with firms experiencing a rapid *rite de passage* from an infant to an exporter, but in others, where technology was complex and marketing was more elaborate, protection had been relatively long lasting in order to provide a longer period of incubation (Kim, 1987).

Seventh, in contrast to Taiwan, which restricted the emergence of large enterprises, the Korean government intentionally created large firms, *chaebols*, as an instrument to bring about the economy of scale in mature technologies and in turn to develop these "strategic industries" and to lead exports and economy. The government helped the capital formation as well as the subsequent diversification of these *chaebols*. Unlike Taiwan, which did not privatize Japanese enterprises confiscated after World War II (Hou and Gee, 1990), the Korean government sold Japanese colonial properties and state-owned enterprises to selected local entrepreneurs with a favorable long-term installment payment during the inflationary period, giving local entrepreneurs a windfall capital gain. Owning all commercial banks, the government helped more significantly the formation of *chaebols* by allocating scarce foreign exchange and preferential financing at the official rates both of which were only one-half the real market rate (Jung, 1989). The state also gave them large import-substitution projects, for which these entrepreneurs imported production technology on turnkey basis with foreign loans, which was guaranteed by the government. Then, foreign debt burdens due to currency devaluations were compensated with increased low interest loans, reducing market risks for these businesses.

Where Korea differs from other developing countries in promoting big businesses was in the discipline its state exercised over these *chaebols* by penalizing poor performers and rewarding only good ones. Good performers were rewarded with new licenses in other industries, leading to further diversification. For those entering risky industries, the government rewarded entrants with licenses in more lucrative sectors, providing a cushion to nurture risky infant industries. In contrast, the government refused to bail out relatively large scale, badly managed, bankrupt firms in otherwise healthy industries, instead selected better managed *chaebols* to take over them (Amsden, 1989). The government as the controller of commercial banks was in a powerful position to punish poorly managed firms by freezing bank credits. As a result, only three of the largest 10 *chaebols* in 1965—Samsung, Lucky-Goldstar, and Ssangyong—remained on the same list 10 years later. Similarly, seven of the largest 10 in 1975 remained on the same list in 1985.

Chaebols' rapid growth and diversification have enormously affected industrial structure and market concentration in Korea. By 1977, 93% of all commodities, or 62% of all shipments, were produced under monopoly, duopoly, or oligopoly conditions, under which the top three producers accounted for more than 60% market share. The 10 largest *chaebols* accounted for 48.1% of GNP in 1980, indicating that Korean industry was even more highly concentrated than that of Taiwan or Japan. However, Chenery et al. (1986) concluded that total factor productivity as well as output grew faster in Korea's highly concentrated economy than in that of almost any other country they studied.

Eighth, the Korean government also set forth exports as something of a life or death struggle to achieve economic growth goals with the small domestic market. Thus, the government had pushed and pulled firms with threats and promises. Firms were granted unrestricted and tariff-free access to imported intermediate inputs. Firms were also granted automatic access to bank loans for working capital for all export activities, even when the domestic money supply was being tightened. These firms also had unrestricted access to foreign capital goods, as mentioned earlier, and were encouraged to integrate vertically in order to sustain international competitiveness. These incentives operated automatically constituted the crux of the Korean system of export promotion. Furthermore, the rationing of longer term bank loans was used as a carrot to draw down to new paths of exporting, encouraging diversification, and to export more than otherwise. These incentives were applied to all exporting firms, but particularly helped large firms with greater organizational, financial, and political leverage to grow even larger.

That the Korean economy was primarily export oriented appears to have affected the fast acquisition of technological capability at least in three ways. First, lump-sum investments for capacity beyond local market size to bring about the economy of scale forced local businesses to acquire technological capability fast in order to maximize capacity utilization through improving international competitiveness. Second, as producers entered the international market, the keen international competition forced them to invest significantly in technological efforts, mainly in learning by doing and reverse engineering, in order to be competitive in both quality and price. Third, informal technical assistance offered by foreign OEM (original equipment manufacture) buyers to ensure Korea-made products meeting their technical specifications provided invaluable help to Korean firms in acquiring necessary capability.

Ninth, while industrial and trade policies affected significantly the emergence and growth of local industries, particularly *chaebols*, science and technology infrastructure played little role in promoting the development of industries with mature technologies. In light of the absence of R&D capability in universities, the government established in 1966 Korea Institute of Science and Technology (KIST) as an integrated technical center and subsequently its spin-offs to meet industry's technical needs. However, the institutes suffered from poor linkages with industry at least through the mid-1970s. In these institutes, most of the overseas trained Korean researchers came from either academic fields or from R&D organizations of highly industrialized countries that undertook advanced research. Expertise was particularly lacking in manufacturing know-how and the development of prototypes, which were in great demand in the early years. Furthermore, Korean researchers could not compete with foreign licensors in supplying detailed blueprints and other manufacturing know-how, as well as being unable to assist industry in solving the problems in the crucial initial stages (Kim, 1989).

Under the aforementioned economic environment, Korean firms developed strategies at the microeconomic level to sustain growth by acquiring technological capability fast, which is the subject to be discussed below.

Microeconomic Behavior of Korean Firms

The microeconomic behavior of Korean firms in acquiring technological capability varies significantly by industry. First, industries using small batch and unit production (hereinafter small batch) such as shipbuilding and machinery focus more on developing capability to design and manufacture products than on setting up production processes, as they usually produce highly differentiated products. As a result, large firms in these industries relied heavily on the formal transfer of foreign technology primarily in the form of foreign licensings and consultancies for the initial erection of production systems and for the design of their products. These industries accounted for a disproportionately large share of Korea's total foreign licensings through the 1970s. These firms have, however, exerted increasingly their own technological efforts to strengthen their own capability and in turn to reduce their reliance on foreign licensors, as they accumulated experience in production and product design. In contrast, small firms in these industries (plus small-sized firms in the electronics and automobile industries, whose large firms deploy large batch operations) resorted primarily to their own imitative efforts to evolve organically over a long period of time, establishing their initial production facilities with primitive technologies developed by themselves and then gradually upgraded their production system and product quality through the imitative reverse engineering of foreign products and processes (Kim and Lee, 1987). Technology diffusion within these industries took rapidly through the mobility of experienced engineers from existing firms to new entrants. Particularly, engineers experienced at state-owned enterprises in the early 1950s have spun-off and played a pivotal role in emerging private firms.

Second, industries using large batch and mass production systems (hereinafter large batch) such as electronics and automobiles produce less differentiated products. Large firms in these industries were also dependent on foreign licensings but to a lesser extent than those with small batch operations. In the advent of an import substitution policy, local firms entered technical assistance agreements with leading foreign firms to import "packaged" technology (including assembly processes, product specifications, production know-how, training local technical personnel, and component parts) so as to establish the initial production system. The immediate technological task was the implementation of transferred foreign technology for the assembly of imported components and parts, whose final products had been tested and proven elsewhere, requiring only engineering (E) efforts. Most late entrants poached experienced personnel from existing firms to set up and begin their operations, resulting in a rapid diffusion of imported technologies. Then, competition stemming from new entrants, together with increasing export promotion, called for indigenous efforts to assimilate foreign technology, thus placing emphasis not only on engineering but also on limited development (D&E). Capabilities acquired through assimilation of foreign imported technology also enabled local firms to develop, through imitative engineering, related products without foreign assistance (Kim, 1980).

Third, industries using continuous processes such as chemicals, cement, paper, and steel produce the least differentiated products in highly capital-intensive processes. Since the final product and the equipment are relatively well known, but the proprietary know-how that lies in the details of the production process can make a significant difference in the productivity of these industries, the initial production system of large chemical, cement and steel plants were established on turnkey basis by Western firms. Initial production capability to operate and maintain such process-oriented plants stemmed largely from extensive training by foreign suppliers before, during, and after the erection of the initial production processes. The local firms, however, used deliberate strategies with considerable efforts to acquire capability, leading to a series of minor improvements with significant productivity increase, and to an increased taking-charge of engineering tasks in the subsequent expansions. A steel mill, for instance, relied completely on the Japanese in the first phase in 1971 for annual capacity of one million tons. But by 1981 capacity had been increased to 8.5 million tons in three expansions that were increasingly under the direction of Korean engineers. The progressive substitution of foreign engineering by local engineering is well indicated by the fall in the cost of foreign project engineering per ton from $6.13 to $0.13 during the same period (Kim, 1987). Unlike other process-oriented industries, local pharmaceutical firms in Korea started as small importers/dealers of packaged finished drugs and later entered the drug manufacturing business by packaging imported bulk drugs. Then, they gradually extended into more intricate operations, first from formulating imported raw materials and later to backward integration into chemical components production. Through this process, they grew in size and technological capability (Kim et al., 1989).

In short, given the large size investment and lack of experience and capability, large firms across industries have resorted heavily to foreign sources in order to ensure quick erection and smooth start-up. In contrast, small firms underwent a long process of imitative reverse engineering, as they lacked financial resources and organizational capability to identify and negotiate with foreign technology suppliers. Nevertheless, both large and small firms deployed deliberate, aggressive technological strategies to acquire their own technological capabilities through exerted efforts to assimilate foreign technologies from the very outset. Korean firms have acquired technological capabilities in piecemeal fashion, as successively more sophisticated capabilities have been acquired and put into practice. The process of acquisition has clearly been one of purposive efforts involving a succession of incremental steps, with production capabilities being developed somewhat in advance of engineering and innovation capabilities. The selectivity of import substitution for the elements of technology has meant continued reliance on imports for at least some elements in almost all industries, but the pattern of imports has continually shifted as local capabilities have replaced foreign ones (Westphal et al., 1985). R&D in the formal sense of the term was not important for Korea during this stage of imitating mature technologies. Industries in fact reversed the sequence of R,D&E: it started with engineering (E) for products and processes imported from abroad, and then progressively evolved into the position of undertaking substantial development (D). But research (R) was not relevant to Korea's industrialization through the 1970s. Several studies conducted in other countries also provide a similar evidence. Studies on Japanese industrial history show that its industries went through a similar pattern (Shishido, 1972; Yakushiji, 1977; Ozawa, 1974). The United States also reversed the sequence of research, development, and engineering, that is, it started with practical technologies imported from Britain, and then slowly evolved into the present position of being afforded substantial research (NAS, 1973). Brazilian and Argentinian firms began manufacturing with assembly and packing operations and eventually extended into more intricate operations. Engineering was the initiating portion of the R,D&E spectrum leading gradually to more advanced development and research efforts (Baranson, 1973; Dahlman and Frischtak, 1990; Katz and Bercovich, 1990).

The rapid acquisition of capability through such a process may be attributed to many factors. The continuous inflow of foreign technology (through formal and informal mechanisms) to learn from, the formation of highly trained human resources who were willing to learn, entrepreneur's "can do" spirit exercised freely under restricted equity participation of multinationals, and the government's orchestral role in directing *chaebols* and selectively allocating resources to them to achieve ambitious growth objectives under the pressure of the competitive international market have worked together for Korea to grow rapidly through the 1970s.

Chaebols particularly played an important role in this process, as they were in the most advantageous position in drawing best quality human resources, in identifying, negotiating, and financing foreign technology transfer, in obtaining business licenses and preferential financing from the government, in applying experience gained in one field of business to another, and in risk-taking in new businesses with existing businesses providing cushion therefor. As a result, by the end of 1970s, Korea had the largest textile plant, the largest plywood plant, the largest shipyard, the largest cement plant, and the largest heavy machinery plant in the world.

CONTEMPORARY NATIONAL SYSTEM FOR INDUSTRIAL INNOVATION

The economic environment for Korea has changed significantly in the 1980s for several reasons. First, the world economy has generally slowed down in the 1980s, affecting particularly outward-looking economies such as Korea. Second, in the wake of rising trade imbalance, North America and Europe have moved toward protectionistic policies, making it increasingly difficult for Korea to sustain export growth in industries that have led Korea's export-oriented strategy in the past. Third, Korea has lost its competitiveness in low wage based labor-intensive industries, as its real wage has risen at the average annual growth rate of 5.8% in the 1960s and 7.5% in the 1970s. Concomitantly, other developing countries with much lower wage rates are rapidly catching up with Korea in these industries. Fourth, advanced countries, particularly Japan, are increasingly reluctant to transfer technology to Korea as Korea attempts to enter industries that have been dominated by these countries. Fifth, Korea was forced to change its copyright and patent laws, preempting the imitative reverse engineering of foreign products. Such an unfavorable environment makes it imperative for Korea to enhance its own technological capabilities in order to innovate independently and to strengthen its bargaining power against foreign technology suppliers. This section focuses on Korea's national system in the 1980s that are designed to cope with these problems.

The Shift of Public Policies

In the face of such an unfavorable environment, the Korean government set out on an economic liberalization program, making a series of policy shifts in an attempt to introduce market mechanisms by reducing government intervention and to undertake a structural change towards the development of relatively more technology-based industries. First, in the face of *chaebols'* increasing economic power and its resultant monopolistic abuses such as creating scarcities, price gouging, and ruining smaller competitors, the government adopted a policy of "economic democratization" to mitigate monopolistic abuses by *chaebols*. The recently introduced Fair Trade Act along the lines of American antitrust legislation includes, among other things, the prohibition of unfair cartel practices and mutual investment among *chaebol's* affiliate companies, sets a ceiling of credit flow going to predesignated *chaebols*, and regulates their vertical and horizontal integration. Nevertheless, *chaebols* grew even bigger and economic concentration increased further: the combined sales of the five largest *chaebols* as percent of GNP increased from 12.8 in 1975 to 35.0 in 1980 and to 52.4 in 1984. Lee et al. (1986) report that the number of *chaebols* designated by the government as dominating their respective markets increased from 105 in 1981 to 216 in 1985, but only 10 were accused of having abused their economic power. Of 1172 applications for vertical or horizontal integrations, only two were rejected by the government.

Second, as part of a financial liberalization program, the government reduced regulation of nonbank financial intermediaries and denationalized commercial banks. The deregulation of nonbank financial intermediaries, many of which had long been controlled by *chaebols*, resulted in their share of total deposit liabilities from 27% in 1980 to 42% in 1984. The denationalization of commercial banks resulted in the fall of their ownership from government hands into those of *chaebols*. With increased financial resources, *chaebols* went further to buy state enterprises that were being privatized and to buy financially troubled firms. In short, the financial liberalization program enhanced the growth of *chaebols*' economic concentration. Thus, Amsden (1989, p. 137) concludes, "in the presence of large concentrations of market power, reliance on the market mechanism for reform appear to produce some perverse results, not only in Latin America but also in the Far East."

Third, since 1983, the government commenced a program to liberalize imports, raising import liberalization ratio (defined as the ratio of the number of unrestricted items to the total) from 51% in 1973 to 85% in 1984 and to 95.2% by 1988. The government also brought down the average tariff rate from 26% in 1984 to 16% in 1988. As a result, imports increased by 20.1% in 1989 compared to 2.8% in export increase, making the domestic market significantly more competitive than before.

Fourth, Korea gradually liberalized its technology transfer policies in the 1980s: Korea needs progressively more sophisticated foreign technologies than in the earlier years in order to sustain its international competitiveness in export-led industrialization. The liberalization raised the share of Korea's 999 industrial subsectors open to foreign investment from 44% in the 1970s to 66% in 1984 and to almost 90% in 1988. Foreign licensings has been completely open for all industries and for all terms and conditions. Table 11.2 presents data on technology transfer through DFI and FL to Korea. Over 48% of total DFI and 67% of royalty payments in FL since 1962 has taken place in the last 5 years (1986–1991), reflecting the liberalized public policy environment as well as the private sector's aggressiveness in acquiring more sophisticated foreign technologies. In the 1960s and 1970s, foreign companies invested in Korea to reap Korea's cheap labor costs. Now foreign companies invest to collaborate with Korean companies in relatively more technology-intensive areas.

Fifth, the government began promoting small and medium-sized industries (SMI), particularly technology-based small firms, to remedy the imbalance between the large and small business sectors. The government established "sanctuaries" for SMIs, designating 205 business territories where neither the large corporations nor their affiliates can intrude. A program called The Compulsory Lending Ratio provides that commercial (nationwide) banks should allow more than 35% of total loans to the SMIs. In addition, local or regional banks have to provide more than 80% of their total loans to these SMIs. The government also took initiatives in establishing the venture capital industry as a way to promote the emergence of technology-based small firms, when the private sector had no interest. Specifically, the government enacted a special law to establish the first venture capital firm, jointly funded by the state and a group of private firms. The government took a further step with the enactment of the Small and Medium Enterprise Formation Act in 1986, leading to the emergence of 12 venture business firms, all jointly invested by the state and private sector.

Sixth, the focus of industrial policy has shifted from the promotion of "strategic" sectors to that of innovation-related activities. That is, in the 1960s and 1970s, special incentives—tax exemptions, custom rebates, access to foreign exchange, and other forms of protection or enhancements—were granted to "strategic" industries in order to make them competitive at a world level. In contrast in the 1980s, the government abolished all industry-specific promotion acts, which had been introduced in the 1970s, and instead legislated a new industrial promotion act that ties all incentives with specific industrial activities such as the promotion of R&D and the development of human resources. However, in the late 1980s, the government designated several high-technology industries for promotion.

Finally, the government, as a way to democratize economy and society, liberalized labor movements that had been tightly controlled in the 1960s and 1970s. Given the opportunity, unionists' demands suddenly exploded disorderly, violently, and in many cases unlawfully, resulting in work stoppages, missed export delivery deadlines, lowered product quality, and wage hike beyond productivity increase. The average company suffered from 19.2 days of labor strikes in 1989, up from 5.3 days in 1987. In 1989 alone, labor disputes cost a total of about $6.2 billion in production as well as $1.36 billion in reduced exports. Korea's real wages have increased a cumulative 62.5% since 1987. When adjusted in U. S. dollar terms, the increase amounts to 91.1%, which is 2.8 times that of productivity (*Business Korea,* 1990). In short, labor disputes are a new variable that affects significantly the nation's efforts to innovate in the 1980s and 1990s.

Under these policy environments, both the government and private sector intensified their technological efforts to sustain international competitiveness. The following sections present various aspects of these technological activities.

Structural Change in R&D Activities

Since many skills and activities required in reverse engineering are also the same in R&D, activities that had been called reverse engineering have easily been transformed into activities called R&D, as Korea approached the technological frontier. Reverse engineering involved activities that sensed the potential needs in market, activities that located knowledge or products that would meet the market needs, and activities that would infuse these two elements into a new project. Reverse engineering also involved purposive search of relevant information, effective interactions among technical members within a project group and with marketing and production departments within the firm, effective interactions with other organizations such as suppliers, customers, local research institutes, and universities, and trial and error in developing a satisfactory result. Skills and activities required in these processes are in fact the same in innovation process in R&D.

Korea's R&D efforts may well be reflected in aggregate R&D investment trends. Table 11.3 shows that the total R&D investment increased from W. 10.6 billion (U S$ 28.6 million) in 1971 to W. 1878 billion (U S$ 2.37 billion) in 1987. Though the Korean economy recorded one of the world's fastest growth rates, R&D expenditure rose faster than GNP, increasing its share of GNP (R&D/GNP) from 0.32 to 1.93% during the same period. A drastic increase was seen in the late 1970s and 1980s, reflecting the shifts of government policy from the promotion of the light/mature industries to that of heavy/technology-based industries and from the assimilation of production technology to a strong drive for technological innovation. However, Korea's total

Table 11.3. Major R&D Indicators in Korea (Billion Won)

	1971	1976	1981	1987
R&D expenditures	10.67	60.90	293.13	1,878.0
Funds from government	7.29	39.18	121.73	383.0
Funds from private sources	3.38	21.72	171.40	1,495.0
Government vs. private	68:32	64:36	42:58	20:80
Manufacturing	NA	16.70	111.77	1,114.7
R&D/sales	NA	0.36	0.67	1.83
Gross national products	3,376	13,881	45,126	97,532
R&D/GNP	0.32	0.44	0.65	1.93
Number of researchers (total)[a]	5,320	11,661	20,718	52,783
Government/public institutions	2,477	3,592	5,065	9,184
Universities	1,918	4,811	8,488	17,495
Private sector	925	3,258	7,165	26,104
R&D exp/researcher (W. 1000)	4,306	5,223	14,149	35,580
Researcher/1000 population	0.08	0.33	0.54	1.27
Number of corporate R&D laboratories	1	12	65	455

[a]The figures do not include research assistants, technicians, and other supporting personnel.
Source: MOST, Science and Technology Annals, various years.

R&D expenditure amounts only to about 54% of that of the General Motors Corporation.

Government statistics indicate that basic research accounted for 16.6%, applied research for 19.6%, and development for the remaining 63.8% of the nation's total R&D expenditures in 1987. The statistics also show that the private sector accounted for 40% of the nation's basic and applied research, while universities accounted for only 28% and public R&D institutes for 32% in the same year (KIRI, 1989). These figures for basic and applied research, particularly the share of the private sector therein, appear to have been much exaggerated, as leading *chaebols* have just started a limited investment in applied research in the field of semiconductors and informatics, let alone basic research.

There has been structural change in R&D investment. The private sector has gradually assumed a larger role in the country's R&D efforts in response partly to increasing international competition and partly to policy environment conducive to private R&D activities. For example, the public sector accounted for 68% of the nation's total R&D expenditure in 1971 but for only 20% in 1987. In short, the government took initiatives in local R&D efforts when the private sector had neither the capability nor market incentives to undertake such activities. But as industrialization progressed and Korea lost its comparative advantage in labor-intensive industries, the private sector gradually took the leading role in local R&D in order to sustain its international competitiveness.

The Korean government has also promoted corporate R&D activities by offering various programs to induce the private sector to set up formal R&D laboratories. The programs include tax incentives and preferential financing to set up new laboratories and exemption from military service obligation for key R&D personnel. Spurred by these programs, the number of corporate R&D laboratories increased from one in

1970 to 122 by 1983 and to 604 in 1987, reflecting the seriousness with which Korean firms are pursuing high-technology development. Although small and medium-sized firms account for over 50% of total number of corporate R&D centers, *chaebols* dominate R&D activities (to be discussed in detail below). Table 11.3 also shows that R&D spending in the manufacturing sector has increased from W.16.7 billion (U S$ 22.25 million) in 1976 to W.1, 114.74 billion (U S$ 1.41 billion) in 1987. Even in constant prices, R&D spending has more than quadrupled during the period. This is faster than the growth of sales: R&D spending as a ratio of total sales increased from 0.39% in 1976 to 1.83% in 1987. The machinery and electronics industry spent over 4% of sales for R&D activities in 1987. However, manufacturer's R&D spending as a proportion of total sales is less than half of that of manufacturers in the United States and Japan.

One of the major mistakes made by the Korean government in developing a national system for industrial innovation has been its underinvestment in higher educational institutions. The number of universities has increased significantly from 69 in 1966 to 100 in 1985. The number of university students has increased from 131,354 to 931,884, with the proportion of high school graduates going on to universities growing from 28.2 to 38.3% during the same period. The environment for quality education and research has, however, deteriorated over this period. For example, the student–professor ratio has retrogressed from 22.6 in 1966 to 35.8 in 1985, making all universities primarily undergraduate teaching-oriented rather than research-oriented. Thus, the "publish or perish" principle is not applied.[1]

As a result, Korea has failed to develop a stock of highly trained scientists and engineers who will be necessary in the 1990s for Korea to sustain its international competitiveness. Compared to education at other levels, graduate education in science and engineering has been much neglected in terms of quantity and quality. The government's recent move to correct this discrepancy has resulted in across-the-board expansion of graduate programs in all universities in an attempt to meet future needs at least in quantity. It will take a great deal of time and effort, however, before such a move can produce quality output.

Realizing the difficulty of reforming old undergraduate teaching-oriented traditions in education, the government adopted a dual system: since all universities, public or private, under the Ministry of Education (MOE) are basically teaching-oriented, the Ministry of Science and Technology (MOST) founded a research-oriented S&T school and plans to have another one soon, establishing a new research tradition in university education.

Consequently, research activities in universities have been relatively underdeveloped. For example, though university R&D expenditures increased significantly from W.572 million ($1.5 million) in 1971 to W.198 billion ($250 million) in 1987, they account for only 5.4% of the nation's total R&D expenditures in 1971 and 10.5% in 1987, while universities account for 33.1% of the nation's R&D manpower and 78.4% of its Ph.D. level R&D manpower in 1987.

The Korean government, however, is on the move in strengthening Korea's basic research capability. For example, the government established a joint semiconductor research institute in a state university, which will be used jointly by other universities and corporate R&D centers. The government recently designated five university R&D centers in 1989 that will receive increased financial assistance in order to strengthen their R&D capabilities. The government is also in the process of enacting a "Basic

Research Promotion Law" as a way to promote the strengthening of basic research capability as one of the nation's top technological goals.

In view of the absence of R&D capability in universities, public R&D institutes have been a backbone of advanced R&D in Korea. They have also been major instruments for national R&D projects, as they are recipients of over 90% of research grants awarded by the government in "new" technology areas.

The Funding Source of Industrial R&D

Table 11.3 shows that the private sector has increasingly assumed a bigger role in the nation's R&D, accounting for 80% in 1987. The Korean government has played a significant role in helping the corporate sector secure funds for R&D activities mainly through three mechanisms: direct R&D subsidy, preferential financing, and tax incentives.

Direct R&D Subsidy

The government had no direct R&D subsidy programs through the 1970s, when Korea relied primarily on light industries for economic development. It was in the 1980s when the government introduced two schemes for direct funding of private R&D: one is "National R&D Projects" (NRP) administered by MOST in "new" technology areas focusing primarily on future problems and the other is "Industrial Base Technology Development Projects" (IBTDP) administered by the Ministry of Trade and Industry (MTI) in "existing" technology areas focusing primarily on current problems. Both schemes designate target technologies and offer direct R&D subsidy to R&D organizations.

Criteria for NRP designation include projects involving "new" (to Korea) technologies with a high risk of failure or with high economic externalities, thus warranting public support. MOST identified several target areas: localization of machinery parts and components, new materials development, semiconductor design, super-mini-computer development, energy conservation technology, localization of nuclear energy fuel, new chemical development, biotechnology development, and basic research in universities. In contrast, MTI undertakes a survey each year to identify urgent R&D projects in industrial firms and provides R&D subsidy (IBTDPs) to research organizations (public, private, and university) to undertake the projects. In 1989, 174 technologies were identified, 146 of which were designated as projects to be funded.

The government's total investment in NRPs, as shown in Table 11.4, increased significantly from W.13.3 billion (U S$ 17.7 million) in 1982 to W.55.0 billion (U S$ 69.4 million) in 1987. The government's R&D subsidy to the private sector through NRPs also increased from W.5.0 billion (U S$ 6.7 million) to W.18.1 billion (U S$ 22.8 million). However, even in the case of 1987, this amounts only to approximately U S$ 61,700 per project (equivalent to about two R&D man years) and U S$ 91,000 per participating firm. For IBTDPs, the government earmarked W.11.5 billion (U S$ 17.2 million) in 1989, about U S$ 117,936 per project. The amounts are not substantial enough to solve critical problems.

Preferential Financing for Corporate R&D

The most important mechanism for funding corporate R&D is preferential financing offered by state-controlled banks and public funds. Through the 1970s, the interest

Table 11.4. National R&D Project Investment (Billion Won)

	1982	1983	1984	1985	1986	1987
Public and industrial projects						
Projects with public leadership	8.2	13.9	16.1	17.3	27.1	28.2
Public R&D institute-industry joint projects	4.6	6.7	4.5	8.1	15.1	18.1
(matched by private sector)	5.0	12.6	9.0	13.2	46.1	49.5
Number of projects	66	106	106	186	296	370
Number of corporates involved	86	131	134	212	240	250
International joint projects	—	—	—	1.6	2.3	2.9
Basic research	—	1.0	0.9	1.5	5.0	5.0
Technical assistance to small/ medium firms	0.4	0.4	0.5	1.0	1.5	—
R&D evaluation projects	—	—	—	0.5	0.6	0.8
Total NRPs[a]	13.3	22.0	22.0	30.0	51.7	55.0

[a]National R&D Project.
Source: Science and Technology Annals (1988).

rate for R&D loans was one of the highest, reflecting the low priority of R&D in government policies. Table 11.5 shows that preferential financing amounted to W.671.6 billion (U S$ 848.0 million) in 1987, accounting for 94.3% of total corporate R&D financing (W.712.4 billion equivalent to U S$ 899.5 million) by the state that year. In contrast, direct R&D subsidy through NRPs and IBTDPs accounts for only 4% and direct investment through venture capital firms for 1.7% of the total. The total amount of public financing (W.712.4 billion) mostly in the form of preferential loans, in fact, accounted for 64% of total R&D expenditure in manufacturing in 1987. In short, the government plays a major role in funding corporate R&D through the allocation of preferential financing.

Table 11.5. Public Source of Funding for Corporate R&D (W. 100 Million)

	1982	1983	1984	1985	1986	1987
Direct R&D subsidy						
NRPs[a]	51	71	51	81	151	185
IBTDPs[a]	—	—	—	—	—	100
Inventor's prototype development	0.5	0.7	1.0	1.0	1.1	1.2
Subtotal	51.5	71.7	52.0	82.0	152.1	286.2
Direct investment through venture capital firms	2	23	31	44	60	122
Preferential financing						
Venture capital firms	221	499	601	827	905	1,000
State-controlled banks	521	879	920	880	1,352	1,730
National investment funds	—	9	33	4	193	525
Industrial development funds	95	64	91	139	195	307
Industrial technology improvement funds	—	—	—	—	316	2,654
Subtotal	837	1,451	1,717	1,942	3,252	6,716
Total	890	1,545	1,800	2,068	3,464	7,124

[a]NRPs, National R&D Projects administered by MOST in "new" technology areas; IBTDPs, Industrial Base Technology Development Projects administered by MTI in "existing" technology areas.
Source: Science and Technology Annals (1988).

Tax Incentives
Tax incentives are a major indirect mechanism in making funds available for corporate R&D. In Korea, tax incentives may be classified into five categories according to objectives associated with such incentives. Most important is tax incentives that are aimed at promoting corporate R&D investment: reduced tariffs on import of R&D equipment and supplies, the deduction of annual noncapital R&D expenditures and human resource development costs from taxable income, accelerated depreciation on industrial R&D facilities, and the exemption of real estate tax on R&D-related properties. The incentives also include a tax reduction scheme, called the Technology Development Reserve Fund, whereby an enterprise can set aside up to 20% (30% for high-technology industries) of profits before tax in any one year to be used for its R&D work in the following 4 years. The private sector did not take advantage of this scheme in early years in the absence of needs for technological activities. Other tax incentives are aimed at reducing the cost of acquiring foreign technology, promoting technology-based small firms, reducing the cost of commercializing locally generated technologies, reducing the cost of introducing new products, and promoting the venture capital industry and venture investment.

Interorganizational Cooperation

Public R&D Institutes–Industry Cooperation
In the 1980s, the government used various means to bring about effective linkages between the public R&D institutes and industry. For example, NRPs and IBTDPs, albeit insignificant in their total monetary value, were used as major means to facilitate public R&D institute–industry cooperation to develop the aforementioned target technologies. The schemes were also aimed at facilitating university involvement in the projects, but have produced only limited results. In some cases, large firms, however, tend to shy away from receiving public support in recent years in order to keep their R&D activities confidential; thus the programs now largely support small and medium-sized firms. Nevertheless, both mechanisms have led to significant joint R&D results. For example, a public R&D institute and four private firms have successfully developed and commercialized a low-density electronics switching system (TDX-1) and are now in the process of designing a high-density system (TDX-10). The same public R&D institute and three semiconductor producers completed jointly designing a 4-mega dynamic RAM (to be elaborated further below).

University–Industry Cooperation
Given the teaching orientation of almost all universities, there is fundamental lack of interplay between the universities and the private sector. This is one of the greatest weaknesses of Korea's national system for industrial innovation. Unlike universities in advanced countries that play a major role in basic as well as applied research that have important bearings on industrial R&D, almost all Korean universities suffer from insufficient facilities and their faculty members have neither the time nor the incentives to undertake serious research. Consequently, universities have lost the confidence of the private sectors, drawing almost no research funds from business circles. The government, as a way to promote university–industry cooperation, opened its NRPs to universities but universities participated in only 6 of 255 projects in 1984, involving

W.194 million, or less than 1% of total NRP expenditure (KAIST, 1986). In short, university–industry R&D cooperation in the formal sense of the term has been insignificant. But informal collaborations in the form of consultancy by individual faculty members are quite prevalent. This may be symptomatic of Korea's position as a catch-up country, where technological tasks are not yet at the frontier.

Intercorporate Cooperation

A major form of intercorporate R&D cooperation is industrial research cooperatives. The technology Development Promotion Act of 1977 promotes the formation of such cooperatives, but it was not until 1982 that the first such cooperative emerged in response to direct R&D subsidies offered by NRPs. Eleven cooperatives emerged in the first year. By mid-1989, 46 industrial research cooperatives were founded involving 986 firms, 759 of which are small and medium-sized firms. The government offers various tax incentives to promote R&D cooperatives. Most existing cooperatives, however, are in name only with little substance. NRPs and IBTDPs are used as instruments to bring about effective intercorporate R&D cooperation by inviting R&D cooperatives to undertake joint research with public R&D institutes, but they neither have adequate R&D facilities nor full-time researchers to take advantage of subsidies to R&D cooperatives.

Technological Strategy of the Private Sector

As *chaebols* have been a driving force in expanding production and exports in the 1960s and 1970s, they also play a pivotal role in developing high-technology industries in Korea in the 1980s. In doing so, *chaebols* have aggressively diversified their sourcing of technologies. First, several major *chaebols* have set up outposts in Silicon Valley in California to "leapfrog" into state-of-the-art technologies by monitoring technological changes and to acquire advanced semiconductor and computer technologies. California also gave another advantage to *chaebols* by solving the most obvious obstacle to Korea's high-technology adventures—lack of experienced scientists and engineers for high-technology industries. California is populated by thousands of top-notch scientists and engineers who are Korean-Americans. Leading *chaebols* have lured away some of the best. Many of them left Korea more than a decade ago, earned Ph.D.s in America's best universities, and rose through the ranks of such leading U. S. concerns as IBM, Fairchild, Intel, and National Semiconductor. The well-financed Korean *chaebols* gave them challenging jobs and attractive compensation packages without their having to leave their American life-styles and with considerable independence. For example, Hyundai's first step on entering industrial electronics was establishing a subsidiary—Modern Electrosystems in Sunnyvale, California—manned mostly by Korean-American holders of Ph.D.s and research veterans of top U. S. semiconductor makers. Samsung also established two subsidiaries in Silicon Valley to undertake R&D activities in microelectronics and biotechnology fields, led by several Ph.D.-holding Korean-Americans, with the rest being native American engineers. Another Korean-owned start-up, ID Focus of Santa Clara, undertook market research and industrial design for Daewoo Electronics and Daewoo Telecom with a staff composed entirely of Americans. Lucky-Goldstar also established two subsidiaries for semiconductor/computer and biotechnology R&D. These subsidiaries are engaged mainly in

R&D activities for mass production in Korea. These outposts in California also serve as an "antenna" for information on research activities in advanced countries and as training posts for scientists and engineers from R&D centers and manufacturing plants in Korea (Kim, 1987).

Second, it is also *chaebols* that have successfully developed ties with multinationals, which provide important inputs in developing Korea's high technologies. Multinationals from advanced countries are looking at Korea to form consortiums for technology cooperation, as Korea is very weak in basic design and basic science but has an international competitive edge in manufacturing and process technologies. Multinationals in high-technology areas such as IBM, Hewlett-Packard, Honeywell, ATT, Monsanto, Hitachi, and Toshiba, have recently established joint ventures with Korean *chaebols* in high-technology industries. *Chaebols* have also entered extensive licensing ties with foreign high technology firms.

Third, *chaebols* developed closer ties with local public R&D institutes in the 1980s than the previous decades. Capabilities available in the public R&D institutes became relevant to *chaebols'* efforts to develop high-technology industries.

Fourth, *chaebols* have also invested aggressively in developing in-house R&D activities in order not only to absorb, assimilate, and adapt imported technologies from their outposts in Silicon Valley and from multinational partners but also to strengthen their own innovative activities. For example, Samsung set up 12 new R&D centers in the country during the 1980s, including Samsung Advanced Institute of Technology and Samsung Integrated R&D center. In 1989 alone, Samsung spent $900 million, Hyundai and Lucky-Goldstar spent $600 million each, and Daewoo spent about $300 million in R&D. Approximately 80% of the investment went for electric and electronics, about 18% for chemicals, and the remaining 2% for other fields. These four largest *chaebols* accounted for most of the nation's total industrial R&D. Leading *chaebols* have relative advantage over small or medium-sized firms in recruiting scarce, high-caliber human resources by offering attractive compensation and the best opportunities for career development among Korean firms. It is these *chaebols* that lure foreign-educated Korean scientists and engineers to their R&D centers in Korea and that recruit the top graduates from Korea's best universities. Leading scientists and engineers in *chaebols'* R&D centers and technology planning posts at headquarters are almost all foreign-educated Ph.D.s often with extensive experience at major multinational firms in advanced countries. *Chaebols* also invest most ambitiously in training to upgrade their human resources.

With the technological network, high-caliber scientific and engineering manpower, and ambitious R&D investment, *chaebols* took a leading role in moving toward the technological frontier. Illustrations in some of high-technology industries such as semiconductors, telecommunications, robotics, computers, and aircraft might shed light on how *chaebols* have managed to acquire technological capabilities to start and grow in these industries in the 1980s.

First, Korea has long participated in the semiconductor industry, involving merely simple packaging processes—bonded assembly processing by wholly owned foreign subsidiaries or joint ventures, with all parts and components imported from the parent companies and reexported back to the consignors. Then, it made a technical leap to the production of VLSI (very large-scale integrated circuit) in 1983. Large *chaebols* started the heavy betting to build a semiconductor industry and the odds were stacked in their favor: a ready access to funds syphoned from cash-cow industries, a motivated and highly skilled work force recruited from abroad, and a low labor–cost advantage. But most important of all, there were a number of distressed small semiconductor companies in the United States that were ready to sell what Korea's *chaebols* needed most—chip designs and processes—in attempts to fuel cash for survival. Samsung Semiconductor's 64K DRAM (dynamic random access memory) technology was licensed from Boisie, Idaho-based Micron Technologies, enabling Samsung to hit the market with 64K DRAM some 18 months after the first Japanese ones became commercially available. Design for Samsung's high-speed CMOS process was picked up for $2.1 million from Zytrex of California. Samsung then turned to Mostek for its 256K DRAM design. Goldstar Semiconductor signed a licensing agreement with Advanced Micron Devices in Silicon Valley for chip designs. ATT, the joint venture partner, was a source of design technology for 1M DRAM. Hyundai turned to Vitelic of San Jose for designs of 1M and 256K DRAM as well as for a variety of SRAM (static) designs. Daewoo, the fourth member of the *chaebol* chip club, got its window on the business in 1985 by injecting some $18 million into the ailing Zymos Corp, acquiring 51% equity. Silicon wafers, the most important input for chipmaking, are produced by two local firms: Korsil, a joint venture with Monsanto, and Lucky Advanced Materials, which licensed wafer processing technology from Siltec of California (now owned by Mitsubishi Metals). In short, *chaebols* relied mainly on foreign firms for design and process technologies.

Another important factor that made Korea possible to take root in the semiconductor industry was export restraints imposed on Japan after trade friction with the United States and the subsequent move upmarket by Japanese firms, opening up opportunities for Korean firms to penetrate the U. S. market. This was also true of consumer electronics and subcompact cars (Porter, 1990).

The chipmaking firms made enormous investment in both production facilities and R&D. By the end of 1988, four semiconductor producers invested over $1.3 billion in new facility investment. Total R&D expenditures increased dramatically, from $13 million in 1983 to $95 million in 1987, accounting for 10 and 22% of the total investment, respectively. Industry's aggressive R&D spending cut down Japan's lead. For example, Korea was just 6 months behind Japan in introducing the 1M DRAM.

But the road ahead was getting bumpier and is even more so today. Texas Instruments filed a suit against Samsung and eight Japanese chipmakers for infringing on its patents for DRAM designs in 1986, while Intel Corp filed a similar suit against Hyundai and its American design suppliers. Both Samsung and Hyundai ended up paying royalties on past and future sales of their memory products. Furthermore, work on the next generation of chips—the 4M DRAM—meant exploring the frontiers of semiconductor technology but also competing neck-in-neck with Japanese and U. S. companies. As the stakes have risen in the chip game, the field of players has grown smaller worldwide, meaning that few, if any, of those left in the game can be counted on to sell state-of-the-art chip design technology to Korean *chaebols*. So, Koreans had to tackle the 4M DRAM design alone. To avoid duplicate research and investment, the government stepped in and designated the R&D of the 4M DRAM as a national project. A public R&D institute played the coordinating role with three *chaebols'* participation. Samsung was the first among three *chaebols* that announced its completion of designing 4M DRAM in 1989 and 16M DRAM in 1990, only a few months after Japan.

Second, electronic switching system (ESS) and optical fiber/cable may be high-technology products in telecommunications. The Korean government (the sole buyer) decided in 1981 to shift public telephone switching from an electrical direct dialing (EDD) system to ESS in 1981. Lacking the technological capability to enter ESS production, three *chaebols* and one medium-sized firm entered foreign collaborations—two in the form of joint venture and the other two by licensing—to locally assemble foreign-designed ESS equipment. While the local producers had progressively increased the local content ratio for ESS, raising as much as 76% for some of the models by 1986, a public R&D institute organized a consortium with the four firms to develop Korea's own rural (low-density) ESS system (Time Division Exchange or TDX-1). Production experience of foreign systems plus high-caliber Korean engineering manpower recruited from abroad blended together. On completion, the design technology was transferred to the participating firms for production. TDX-1 has replaced a Swedish model since 1987. The consortium produced an improved version (TDX-1B) and exported it to the Philippines. At the same time, the consortium is undertaking a project to develop a high-density system (TDX-10) for urban use.

When Corning Glass refused to transfer optical fiber production technology to Korea in 1977, two *chaebol* copper cable producers entered a joint R&D project with a public R&D institute. After 7 years of R&D, the locally developed optical cable was tested successfully on a 35-km route in 1983. Although this local effort eventually grounded to a halt due mainly to slow progress in R&D, it nonetheless helped local firms gain bargaining power in acquiring foreign technology on favorable terms. Four *chaebols* entered licensing agreements with multinationals in 1984: Samsung with IT&T, Goldstar with AT&T, Daewoo with Canada's Northern Telecom, and Taihan with Japan's Sumitomo.

Third, Korea took an independent, reverse engineering route in developing robotics in early years. Again four leading *chaebols*—Hyundai, Samsung, Goldstar, and Daewoo—have dominated the industry. The first batch of welding robotics for *chaebol's* automobile assemblers, shipbuilders, and aerospace firms were all imported. The first batch of automatic part inserting robotics for largest electronics firms were all imported. But the users of these imported robotics were active assimilators of imported technology. Based on operating experience of foreign models and capability gained from in-house R&D efforts, they developed imitative models first to meet the subsequent needs within the plant and then to meet increasing demand within their *chaebol* groups. In contrast, some smaller *chaebols* recently entered the industry as a local dealer of foreign models as a way to gain experience. Nevertheless, these *chaebols* stepped up their R&D efforts (over $28 million in 1987 alone) in the robotics industry with the aim of cashing in on fast-growing demand both at home and abroad, giving rapid rise to technological capability and leading to the introduction of their own models. For example, Daewoo's five-axes multiple process NOVA-10 introduced in 1984 had both quality and price competitiveness to export to the United States. As in other *chaebols,* U. S.-trained Ph.D.s played a central role in initiating and expanding the robotics business at Daewoo. The government is also mapping out a comprehensive program to boost automation-related industries such as robotics production. However, robotics technology in Korea is still confined mainly to the production of robotics for simply repetitive welding works, making it difficult to turn out highly integrated products. In response to the need for more sophisticated robotics, *chaebols* such as Hyundai and Daewoo are now pursuing technology tieups with multinationals, while public R&D institutes are undertaking advanced research in areas such as artificial intelligence, vision-sensor, six-axes robotics, and a voice-recognition system.

Fourth, the computer industry has also undergone a similar route in Korea. *Chaebols* with experience in consumer electronics entered the industry through imitative reverse engineering, leading to the introduction of IBM PC (personal computer) clones. These *chaebols* rapidly expanded their businesses as OEM (original equipment manufacture) producers. Smaller *chaebols* without consumer electronics background entered first as a local dealer/service agent as a way to accumulate experience in computer technology and gradually backward integrated to the design and production of their own models (Kim et al., 1987). With a good grip on the chip business and microcomputers, Korea now tackles another national project: the design and manufacture of Korea's own super-minicomputers. Again, a team of software and hardware engineers from a public R&D institute and a number of *chaebols* have been formed to work on the project. The basic idea of the super-mini project is to buy and transplant foreign technology, in this case a super-mini design from one of the U. S. venture firms, and to progressively localize design capability.

Fifth, *chaebols* also dominate the aircraft industry. Korea's rudimentary experience in aircraft technology stemmed largely from international subcontracting and maintenance. Boeing and other U. S. aircraft makers that used Japan as a manufacturing base for bodies, parts, and components have shifted their sourcing to Korea in the facing of rising Yen. Three *chaebols* entered into close relations with U. S. manufacturers. For instance, Daewoo has supplied outer fuselages, side cockpit panels, and central fins to General Dynamics. Daewoo has also supplied upper decks for Boeing 747s and inspar wing ribs, and wing part components for Boeing-737s and wing and tail parts for a new U. S. fighter to Lockheed. Korean Air, which has accumulated aircraft technology from maintenance service of such U. S. military fighters as the F-15, F-16, and F-5 Phantom, has supplied wing tip extensions for Boeing and sheet metal assemblies for McDonnel-Douglas. In addition, Daewoo entered a joint venture with Sikorsky and Samsung exchanged a memorandum of understanding with Bell to produce helicopters in Korea for Korean military and for exports under OEM basis. These *chaebols* will participate in manufacturing the next mainstay fighters for the Korean Air Force. The U. S. government proposed that Korea buys 12 fighters as finished airplanes, produces 36 through kit assembly, and manufactures the remainder under license. Through this process, Korea hopes to acquire enough technological capability to develop its own aircraft. *Chaebols* have intensified their R&D efforts in order to assimilate foreign technologies. The Korean government also takes an active role in promoting the industry: the government legislated in 1987 Aerospace Industry Development Law and established Aerospace Technology Institute as a way to support R&D activities of the private sector.

In short, *chaebols* have deliberately taken a path—from reverse engineering to international tieups and to own research and development—to expedite and cost-minimize in acquiring technological capability in high-technology industries. When technological tasks were relatively simple, they opted to reverse-engineer foreign products. The robotics and computer industries in early years may be good examples. When technological tasks were beyond their capacity, they quickly turned to multinationals, which were willing to share technologies with Korea for various reasons. The semiconductor, electronic switching, aircraft industries in the beginning, and the robotics, computer, and optical fiber industries in the later stage may fit this case. When Korea approached the technological frontier, where few multinationals were willing to share their technologies, Korea consolidated its efforts to develop its own products. The recent development of the 4M DRAM may be a case in point. During this process, public R&D institutes have played a pivotal role in giving rise to the bargaining power of *chaebols* in acquiring foreign technologies and in coordinating joint research efforts with *chaebols*.

Although a complaint is mounting that *chaebols* dominate high-technology industries like they did in many other sectors in the past, this dominance is considered inevitable at the initial stage owing to the great deal of risk and investment and to the absence of research-oriented universities that may incubate technical entrepreneurs who might start small technology-based venture firms. There are, however, some indications that a few venture businesses have recently appeared in high-technology fields. Expanding investment in basic research at universities, increased availability of venture capital, increasingly conducive economic environment created by the government, and the fact that *chaebols* cannot dominate a wide range of application technologies will hopefully lead to a strong appearance of small-sized venture businesses in Korea in the near future.

Other Issues

There are two other issues worth covering here: the effects of military R&D on industrial innovation and the formation of hi-tech "valleys." First, given the threat of hostilities with North Korea, national security has been one of the major concerns in Korea, making the home market unusually sophisticated and demanding. As a way to mitigate the effects of constraints imposed by foreign weapon suppliers on national security and to ensure military independence, the Korean government in the 1970s launched an ambitious program to build up local capability to develop weaponry. Its budget size was, however, $114 million in 1988, only 0.2% of that in the United States and 4% of that in France. The percent of military R&D budget to total military expenditures was only 1.5% in Korea compared to over 10% in France and the United States. The nature of projects, however, has been kept strictly confidential. And R&D has been undertaken almost strictly within the military: only 1.48% of military R&D budget was allotted to universities in 1988 (NDC, 1990). Such isolated military R&D activities appear to have suffered from lack of interaction with the rest of Korea's S&T infrastructure and also from competitive disadvantages vis-à-vis public and corporate R&D institutes in recruting highly qualified scientists and engineers. As a result, the ambitious program has been drastically scaled down in the 1980s. The private sector is involved in manufacturing some traditional weaponry, but the spillover effects of such operations on industrial innovation appear to have been quite negligible.

Second, the government created two high-technology centers: Seoul Science Park and Daeduck Science Town, 200 km south of Seoul. Seoul Science Park started with three R&D institutes and three economic research institutes but has failed to attract private R&D centers in the vicinity. Furthermore, one of the three R&D institutes has relocated, while the second one will be relocated within a few years.

In contrast, the government has actively promoted the development of Daeduck Science Town, where it has located nine public R&D institutes and three tertiary educational institutions and attracted four private R&D laboratories. About 20 major firms plan to locate their central R&D laboratories in the Town within a few years, making the Town the first high-technology "valley" in the country.

In addition, there are signs in different parts of the country that corporate R&D laboratories in some fields tend to cluster together in the same vicinity. For example, semiconductor R&D centers and biotechnology R&D centers cluster near Seoul, telecommunications R&D laboratories near Kumi, which is known as an electronics industrial zone, and new materials R&D centers near Pohang, where POSCO has its first steel plant and its own research-oriented private university. It may be premature to regard them as high-technology "valleys" at this point but they appear to be in the incubation stage.

CONCLUDING REMARKS

Korea has been transformed from a subsistent agricultural economy into a newly industrializing one in a very short span of time. During the transformation process, Korea's economic environment has changed drastically. Protectionist moves in North America and Europe are making it increasingly difficult for Korea to sustain export growth in the industries that have led Korea's outward-looking economic strategy in the past. These countries have also put pressure on Korea to liberalize its domestic economy, forcing Korea to compete against multinationals not only in international markets but also in its domestic market. At the same time, Korea's competitiveness in labor-intensive industries had been rapidly eroded by rising real wages. Other developing countries with much lower wage rates are rapidly catching up with Korea in labor-intensive industries. Furthermore, industrialized countries, particularly Japan, are increasingly reluctant to transfer technology to Korea (Kim, 1988).

The rapid changes in economic environment have not given Korea enough time to make and consolidate a major shift in designing an appropriate national system for industrial innovation. The government has introduced in the 1980s various new policy instruments in an attempt to develop a new national system, which is yet at the formative stage. Furthermore, a consensus on the method of redesigning the institutional system and on drastic increases in resource allocation for R&D expenditure is difficult to be made. At the same time, the traditionally teaching-oriented universities could not easily be transformed into research-oriented institutions, giving gloomy implications for generating high-caliber scientists and engineers, let alone university–industry cooperation. If the formation of skilled workers in the 1950s were an important leading indicator for the subsequent development in the 1960s and 1970s, then the poor health of universities and the resultant lack of high-caliber human resources in the 1980s may be a critical warning signal for the 1990s and on.

The government took the initiatives in the 1960s and 1970s in directing and driving the private sector to achieve growth by fostering the formation of large *chaebols* through the allocation of projects and resources. This scheme appears to have helped in bringing about an economy of scale in the stable, mature industries and achieving outward-looking objectives in the past. At the same time, however, the same scheme gave rise to highly concentrated economic power within a limited number of *chaebols* and stifled the growth of small and medium-sized industries. These *chaebols* may be in an advantageous position in the 1980s in developing such industries as semiconductors and aircraft, which require a large investment. The highly concentrated market structure has, however, curtailed the market dynamism. Furthermore, highly bureaucratized large *chaebols,* which effectively functioned in the stable industries in the past, now find themselves rigid and slow in adapting to rapid changes in technology and market. Some *chaebols* have, nonetheless, managed to build up their capability to cope with the changing economic environment. They began investing heavily in R&D to give rise to existing capability in order to design and manufacture "next stage" products. But most others are highly dependent on foreign sources for new technologies.

Another problem in Korean economy is the vital missing link in the presence of related and especially supporting industries. *Chaebols* have largely focused their efforts in end products with heavy dependence on imported parts and machinery. The problem, as Porter (1990) notes, is not the cost of inputs but the effects on the innovation process. Without fluid supports from capable domestic suppliers, Korean firms will remain behind in product and process innovations.

Korea is, indeed, at the crossroad. While the government and private sector exert their efforts to redesign the national system for industrial innovation and in turn to sustain the nation's economic growth by strengthening Korea's technological capabilities, there are several critical signs, as mentioned above, that have led Korea to a serious economic crisis since 1989 and that may impede its future development.

NOTE

1. Exceptions are Korea Advanced Institute of Science and Technology (KAIST), a graduate school of applied science, and Korea Institute of Technology (KIT), an undergraduate school within the KAIST system, both of which are under the jurisdiction of MOST. All other universities and colleges are under the jurisdiction of the Ministry of Education (MOE). Another exception is POSTEC, a new research-oriented school of science and technology established by Pohang Iron and Steel Corporation (POSCO), one of the most cost-competitive steel companies in the world.

REFERENCES

Amsden, A. H. (1989). *Asia's Next Giant: South Korea and Late Industrialization.* New York: Oxford University Press.

Baranson, J. (1973). "The Drive Toward Technological Self-Sufficiency in Developing Countries." Paper presented at the Conference on Latin America-United States Economic Interactions: Conflict, Accommodation, and Policies for the Future, at the University of Texas at Austin, March 19.

Business Korea (1990). "Balancing Workers' Interests against the Nations." February, p. 31.

Chenery, H. Robinson, S., and Syrquin, M. (1986). *Industrialization and Growth: A Comparative Study.* New York: Oxford University Press.

Dahlman, C. J., and Frischtak, C. R. (1990). "National Systems Supporting Technical Advanced in Industry: The Brazilian Experience." Paper presented at National Technical System Conference at Stanford University, October 18–19.

Harbison, Frederick, H., and Myers, C. S. (1964). *Education, Manpower, and Economic Growth.* New York: McGraw-Hill.

Hou, C. M., and Gee, S. (1990). "National Systems Supporting Technical Advance in Industry—The Case of Taiwan." Paper presented at National Technical System Conference at Stanford University on October 18–19.

Jung, K. H. (1989). "Business-Government Relations in Korea." In K. H. Chung and H. C. Lee (eds.), *Korean Managerial Dynamics*, 11–26. New York: Praeger.

KAIST (Korea Advanced Institute of Science and Technology) (1986). *Urinara Kwahak Gisul Gebal Systemeo Jeonge Gwajungkwa Jeonmang.* (A Study on Science and Technology Development System in Korea and Its Future Direction). Seoul: Ministry of Science and Technology.

Katz, J., and Bercovich, N. A. (1990). "Science, Technology and Socioeconomic Re-structuring: The Case of Argentina." Paper presented at National Technical System Conference at Stanford University, October 18–19.

KEB (Korea Exchange Bank) (1987). "Direct Foreign Investment in Korea." *Monthly Review* October, 3–13.

Kim, K. S., and Roemer, M. (1979). *Growth and Structural Transformation.* Cambridge, MA: Council on East Asian Studies, Harvard University Press.

Kim, L. (1980). "Stages of Development of Industrial Technology in a LDC: A Model." *Research Policy* 9:254–77.

Kim, L. (1987). *Technological Transformation in Korea: Progress Achieved and Problems Ahead.* Helsinki: World Institute for Development Economic Research, mimeo.

Kim, L. (1988). "Korea's Acquisition of Technological Capability for Internationalization: Macro and Micro Factors." *Business Review* 22(1): 183–97.

Kim, L. (1989). "Science and Technology Policies for Industrialization in Korea." In J. Suh (ed.), *Strategies for Industrial Development*, 255–272. Kuala Lumpur: Asia and Pacific Development Council.

Kim , L., and Kim, Y. (1985). "Innovation in a Newly Industrializing Country: A Multiple Discriminant Analysis." *Management Science* 31(3): 312–22.

Kim, L., and Lee, H. (1987). "Patterns of Technological Change in a Rapidly Developing Country: A Synthesis." *Technovation* 6:261–76.

Kim, L., Lee, J. and Lee, J. (1987) "Korea's Entry into the Computer Industry and Its Acquisition of Technological Capability." *Technovation* 6(4):277–93.

Kim, Y., Kim, L., and Lee, J. (1989). "Innovation Strategy of Local Pharmaceutical Firms in Korea: A Multivariate Analysis." *Technology Analysis & Strategic Management* 1(1):29–44.

KIRI (Korea Industrial Research Institutes) (1989). *Major Indicators of Industrial Technology.* Seoul: KIRI.

Lee, K. U., Urata, S. and Choi, I. (1986). "Recent Development in Industrial Organizational Issues in Korea." Mimeo, Korea Development Institute and the World Bank, Washington, D.C.

Mason, E. S., Perkins, D. H. Kim, K. S. Cole, D. C. and Kim, M. J. (1980). *The Economic and Social Modernization of the Republic of Korea.* Cambridge, MA: Council on East Asian Studies, Harvard University Press.

McGinn, N. F., Snodgrass, D. R., Kim, Y. B., Kim, S. B. and Kim, Q.-Y. (1980). *Education and Development in Korea.* Cambridge, MA: Council on East Asian Studies, Harvard University Press.

MOST (Ministry of Science and Technology) (various years). *Science and Technology Annals.* Seoul: MOST

NAS (National Academy of Sciences) (1973). *U. S. International Firms and R,D&E in Developing Countries.* Washington, D. C.: National Academy of Sciences.

NDC (National Defence College) (1990). *Gukbang yungu gebalkwa mingan yungu gebaleo gyunge ganghwa bangane kwanhan yungu.* (A proposal to strengthen the link between military and industrial R&D). Seoul: KIST Center for Science and Technology Policy.

Ozawa, T. (1974). *Japan's Technological Challenge to the West, 1950–1974: Motivation and Accomplishment.* Cambridge, MA: MIT Press.

Porter, M. E. (1990). *The Competitive Advantage of Nations.* New York: The Free Press.

Shishido, T. (1972). *Japanese Policies for Science and Technology.* Tokyo: Nikko Research Center, mimeo.

Westphal, L. E., Kim, L. and Dahlman, C. (1985). "Reflections on the Republic of Korea's Acquisition of Technological Capability." In N. Rosenberg and C. Frischtak (eds.), *International Technology Transfer: Concepts, Measures, and Comparisons* 167–221. New York: Praeger.

Yakushiji, T. (1977). *Dynamics of Policy Interventions: Government and Automobile Industry in Japan, 1900–1960.* Ph.D. dissertation, MIT, Cambridge, MA.

[11]

Industry and Innovation, Volume 4, Number 2, December 1997

THE DYNAMICS OF R&D IN INDUSTRIAL DEVELOPMENT

LESSONS FROM THE KOREAN EXPERIENCE

LINSU KIM AND GIHONG YI

Industrial development is a process of acquiring technological capabilities and translating them into product and process innovations in the course of continuous technological change. In advanced countries, "learning by research" by corporations, universities, and government research institutes (GRIs) plays a dominant role in expanding the technological frontier. In developing countries, by contrast, "learning by doing" and imitative reverse-engineering by corporations, with limited assistance from universities and GRIs, is a dominant pattern of accumulating technological capabilities. A few newly industrializing economies (NIEs) have made the rapid transition from "learning by doing" to "learning by research". Korea must be counted as one of them.

Indeed, Korea has been transformed from a subsistent agrarian economy into a newly industrialized one in the space of the past three decades. Beginning in 1962, the Korean economy grew at an average annual growth rate of almost 8 percent, raising GNP per capita in current prices from $87 per year in 1962 to $10,007 per year in 1995. Exports increased from a mere $40 million to $125 billion during the same period. In the mid-1960s, Korea began exporting textiles, apparels, toys, wigs, plywood, and other labor-intensive mature products. Ten years later, ships, steel, consumer electronics, and construction services from Korea challenged established suppliers from the industrially advanced countries. By the mid-1980s, computers, semiconductor memory chips, video cassette recorders, electronic switching systems, automobiles, industrial plants and other technology- and knowledge-intensive products were added to the list of Korea's major export items.

In the mid-1990s, Korea is working on the next generation products such as multi-media electronics, high density television, personal communication systems and a new type of nuclear breeder. By this time, Korea had approached the world technological frontier: it ranked second in the world in shipbuilding and consumer electronics, third in semiconductor memory chips, fifth in textiles, chemical fibers, petrochemicals, and electronics, and sixth in automobiles and iron and steel (KDB 1994). Such phenomenal growth and structural change in exports may be attributed to many social, economic, and technological factors, but most important of all may be technological learning (Kim 1997a).

In this process of industrial development, Korean firms, particularly *chaebols* (the Korean version of Japanese *zaibatsu*), have transformed themselves from mere imitators of mature technologies to contenders in some of cutting-edge technologies

1366-2716/97/020167-16 © 1997 Carax Publishing Ltd

in three decades. The Korean government has also made significant investment in universities and GRIs in the hope that the R&D community (referring primarily to universities and GRIs in this paper) would play a vital role in helping industries to strengthen their technological capabilities and in turn boost their international competitiveness.

This paper addresses the question: How have Korean firms upgraded their technological capabilities over time? Has the R&D community been effective in helping the industry? If not, why not? How does its role evolve over time as industrialization progresses? These questions address vital issues facing many developing countries today, as well as industries in advanced countries which are seeking to upgrade their technological capabilities.

This article frames its discussion in the context of a dynamic model of R&D activity in a newly industrializing country, namely Korea. Such a model places as much emphasis on the role of universities and GRIs, particularly in early years, as on the research role of private firms, which emerge as the premier bodies responsible for R&D at the later stages of industrial development. The paper utilizes a three-stage model of development, with public sector institutions such as GRIs and private sector firms playing different roles in R&D activity in each period. The framework is shown in Table 1. The paper examines the evolving role of R&D in corporations, universities, and GRIs and their interactions, within each of the three stages—in transition from the early stage, to the intermediate stage, and to the knowledge-intensive stage in Korea. It then discusses the implications of the Korean experience for other developing countries.

IMITATIVE REVERSE-ENGINEERING IN EARLY STAGE INDUSTRIALIZATION

In the early years of industrialization, firms in developing countries import or imitate mature technologies, whose products and markets have already been well tested in the advanced countries (Pack and Westphal 1986). Technology tends to be readily available in machine-embodied form (Kim and Kim 1985). Large firms acquire production technologies mainly through turnkey plants or foreign license. Given the scale of large investment required, large local firms at this early stage are highly motivated to look to experienced foreign firms to ensure swift construction and smooth start-up. In contrast, small firms took an imitative approach, reverse-engineering foreign products and evolving organically over a long period time. Both large and small firms deployed deliberate and aggressive strategies to assimilate foreign technologies (Kim and Lee 1987). Thus, R&D in the sense of generating new knowledge was not needed; only limited engineering development (ED) was necessary. Such imitative reverse engineering was possible only because Korea had a good stock of well-trained human resources. Some developing countries attained rapid growth in elementary education. But what was unique in Korea was the well-balanced expansion in all levels of education early enough to support its economic development. In the 1960s, using data from the late 1950s for 73 developing countries, Harbison and Myers (1964) found three nations—Korea, Taiwan, and Yugoslavia—with levels of educational achievements far above what would be expected, given their levels of

TABLE 1: THE EVOLUTION OF R&D ACTIVITIES IN KOREA

	Initial Stage	Intermediate Stage	Knowledge-intensive stage
Business R&D	—Little R&D investment —Imitative reverse-engineering —Limited engineering (E)	—Formative stage —Advanced reverse-engineering —Development and engineering (D&E)	—Dominant role in the nation's R&D —Globalization of R&D —Research, development, and engineering (RD&E)
University R&D	—Minimal role —Undergraduate teaching-oriented	—Formative stage —Informal links with industry	—Basic research being strengthened —Stronger formal links with industry
Government research institute R&D	—Strengthening industry's bargaining power in technology transfer —Training experienced researchers —Reverse-engineering of advanced technologies —Leading role in the nation's R&D	—Expansion of GRI network —Incubating experienced researchers —Leading role in national R&D projects	—Leading role in national R&D projects —Technical support for SMEs

economic development. Two of these are the countries that made phenomenal industrial development in the subsequent years.

During this period, universities played little role in helping industry in Korea. They remained primarily as undergraduate teaching-oriented institutions, undertaking little research. As shown in Table 2, university R&D expenditure was a mere W400 million (US$1.3 million) in 1970, accounting for only 3.5 percent of the nation's total R&D expenditures. This was insignificant compared with the proportion of the nation's researchers who were affiliated with universities in the same year (over 35 percent).

In the absence of research in universities, the government took the initiative in establishing a GRI—the Korea Institute of Science and Technology (KIST)—by recruiting overseas-trained Korean scientists and engineers. As shown in Table 2, GRIs accounted for 83.9 percent of the nation's total R&D expenditures and 43.7 percent of the nation's pool of researchers in 1970, reflecting their dominant role in R&D activities in Korea. This figure also shows that the number of researchers at KIST and small GRIs exceeded that of universities through 1970.

GRIs, however, faced numerous obstacles in the 1970s. For example, they suffered from poor linkages with industry. Most Korean scientists and engineers recruited by

TABLE 2: RESEARCH AND DEVELOPMENT INVESTMENT IN KOREA, 1965–1995

	1965	1970	1975	1980	1985	1990	1995
R&D expenditures	2.1	10.5	42.7	282.5	1,237.1	3,349.9	9,440.6
Universities		0.4	2.2	25.9	118.8	244.3	770.9
GRIs		8.9	28.1	104.5	367.2	731.0	1,766.7
Private sector	0.2	1.3	12.3	81.4	751.0	2,374.5	6,903.0
Public R&D vs private	61:39	97:03	71:29	64:36	25:75	19:81	19:81
R&D/GNP	0.26	0.38	0.42	0.77	1.58	1.95	2.69
Number of researchers (total)*	2,135	5,628	10,275	18,434	41,473	70,503	128,315
Universities	352	2,011	4,534	8,695	14,935	21,332	44,683
GRIs	1,671	2,458	3,086	4,598	7,542	10,434	15,007
Private sector	112	1,159	2,655	5,141	18,996	38,737	68,625
R&D exp/researcher (W.1,000)	967	1,874	4,152	15,325	27,853	47,514	73,574
Researcher/10,000 population	0.7	1.7	2.9	4.8	10.1	16.4	28.6
Number of corporate R&D centers	0	1**	12	54	183	966	2,270

Unit: billion Won.
Note: * The figures do not include research assistants, technicians, and other supporting personnel.
** For 1971.
Source: Ministry of Science and Technology, Korea.

KIST came from either academic institutions or R&D organizations that undertook advanced research. There was no demand from industries for the kind of service that KIST could offer. Expertise was particularly lacking in manufacturing know-how and the development of prototypes, which were in great demand in the early years. Furthermore, KIST could not compete against foreign firms in supplying detailed blueprints and other manufacturing know-how, as well as being unable to assist industries in solving teething problems in the crucial initial stages of production.

This does not necessarily mean that KIST did not produce any significant outcomes. In fact, researchers at KIST developed several meaningful results. Skeptical about its engineering and production capability, the private sector, however, was reluctant to commercialize these results. The private sector preferred turnkey transplants or technology licensing from experienced foreign firms to technology transfer from the inexperienced local GRIs such as KIST. (Large projects such as fertilizer, chemical, and cement plants relied completely on turnkey basis, while consumer electronics resorted to foreign licensing.)

After several failed attempts to transfer the research results to the private sector, KIST founded the Korea Technology Advancement Corporation (K-TAC), a for-profit venture, to commercialize them. K-TAC, in turn, established several production subsidiaries and served as a holding company to finance and supervise them. Products commercialized successfully through this mechanism include Vitavax (a raw material for an insecticide), HOP (an intermediate material for an insecticide), and metallic powder (KIST 1994).

Despite its problems, KIST made important contributions to industrial development in Korea. One of the most important roles played by KIST at this stage was in helping industries strengthen their bargaining power in acquiring foreign technology. In an attempt to develop an effective linkage between KIST and the private sector, the government coerced large firms to undertake joint research with KIST. Such joint research provided opportunities for some firms to acquire sufficient prior knowledge about technology to be imported. This enabled them to identify prospective technology suppliers and to enhance their bargaining power in negotiating technology transfer arrangements. Once imported, such joint research provided a platform on which the firms could assimilate and adapt technology rapidly. In other words, during the early years of industrialization the private firms entered joint research programs not so much to obtain significant research outcomes as to gain initial knowledge about technology which they were interested in acquiring.

For example, when black and white TV sets reached a rapidly declining stage in the export market, the color TV set became the next target product for Korean firms to sustain ever increasing exports. No foreign color TV producer was willing to license technology to Korean producers and help them invade the US market again, as they did with black and white TV sets. Three major TV set producers therefore jointly entered a research contract with KIST in order to gain sufficient knowledge and experience in color TV technology. Experience gained from black and white receivers and learning from the joint research made it possible for local firms to strengthen their bargaining power and brought the royalty rate significantly down in licensing core patents held by RCA in 1974, enabling them to enter color TV set production and build up exports. In his study of Latin American countries, Utterback (1975) also concludes that the major role of public R&D centers in developing countries is to facilitate and lubricate foreign technology transfer by assisting in the private sector's acquisition of foreign technology, formally or informally.

KIST also played a significant role in transferring technology to industry through reverse-engineering of foreign technology—an activity which was beyond the capacity of Korean industry at the time. A case in point involves polyester film production for use in cassette tapes. When a Japanese company rejected Korean requests for technology transfer, for fear of losing its market in Korea, a Korean chemical firm in collaboration with KIST successfully undertook a reverse engineering task to invent around the production technology. At that time, Korea recognized process patents but not product patents. KIST had no sooner invented around the technology when the Japanese company offered a technology transfer arrangement, which the Korean government rejected in order to protect the Korea-developed technology. Korea now is the world's major supplier for audio and video cassette tapes (Kim 1991).

Another important economic effect of local technology development by GRIs was a drastic cut in the prices of imported technologies and materials. For example, no sooner had KIST successfully developed Betamethasone in 1979 than its import price was halved, from 30,000 won per kg to 15,000 won. When the GRI developed

Rifamycine in 1982, again its import price dropped by more than half, from 1 million won per kg to 0.45 million won (KIST 1994).

The most important but unintended role the GRIs played during the early years of industrialization in Korea was the production of experienced researchers. When the private sector was reluctant to invest in R&D, the GRIs generated a large number of experienced researchers, who later spun-off to new GRIs and emerging corporate R&D centers in the 1980s. For example, over 2,800 experienced researchers left KIST to pursue careers elsewhere: 420 to private R&D centers, 784 to universities, 1,594 to newly established GRIs (KIST 1994).

EMERGENCE OF R&D IN INTERMEDIATE STAGE INDUSTRIALIZATION

As the industrialization process unfolds and firms master manufacturing competencies in standardized, low-cost products, they seek to upgrade their capabilities and the value-added of their production in the face of increasing wages and emerging competitive threats from the second-tier developing countries. While their own R&D activities play a central role in achieving such upgrading, the R&D community (e.g. GRIs and universities) also plays a supplementary role for rapid technological upgrading. We shall see how in the case of Korea industry intensified its in-house R&D and the R&D community came to play a role rather different from that of the early stage in Korea.

The government at this time launched various programs to induce the private sector to set up formal R&D laboratories. These include tax incentives and preferential financing for setting up new laboratories and exemption from military service obligation for key R&D personnel. Spurred by these programs and in response to increasing market competition, the number of corporate R&D laboratories increased from only one in 1970 to nearly 1000 (966) by 1990, reflecting the seriousness with which Korean firms were pursuing more technology-intensive development. The newly established corporate R&D centers were able to draw on the existing GRIs for most of their experienced researchers, as well as on Korean graduates working abroad.

Consequently, this period saw significant structural change in R&D investment. The government took initiatives in Korea's R&D efforts in the early stage when the private sector had neither the capability nor market incentives to undertake such activities. But as industrialization progressed and Korea lost its comparative advantage in labor-intensive industries, the private sector rapidly took the leading role in local R&D in order to sustain its international competitiveness. For instance, the average annual growth rate of the nation's R&D investment as a proportion GDP over the decade of the 1980s (1981-1991) was the highest in the world, at 24.2 percent—compared to 22.3 percent in Singapore, 15.8 percent in Taiwan, 11.4 percent in Spain, and 7.4 percent in Japan. The average annual growth rate of business R&D per GDP was also the highest in Korea (31.6 percent) compared to 23.8 percent in Singapore, 16.5 percent in Taiwan, 14.0 percent in Spain, and 8.8 percent in Japan (DIST 1994). Consequently, the share of the private sector in the nation's total R&D

expenditure rose from a mere 3 percent in 1970 to 75 percent by 1985. This was a dramatic transition by any standard.

Other important indicators of Korea's rapid growth in industrial R&D were patent registrations in Korea and abroad. Domestic patent growth rate has significantly increased from mere 48 percent in 14 years (1965–1978) to almost triple in the next 11 years (1979–1989), reflecting the increasing importance of intellectual property rights in the face of fading reverse engineering.

Several cases demonstrate the substantial in-house R&D in Korean firms. For example, when Japanese and US producers were reluctant to transfer their patented technologies related to videocassette recorders and microwave ovens, Korean firms reinvented them by intensifying in-house R&D, giving them stronger bargaining power in licensing technology from the patent holders. In this case, foreign licensing cleared a path to the export market rather than to technology. Magaziner and Patinkin (1989) recount how Samsung developed the microwave oven and became one of the world's leading producers.

While companies themselves started to pick up more of the R&D burden in the 1970s and 1980s, the government also took steps to ensure that there was a sound technological capability in both universities and GRIs. Frustrated in its efforts to reform the undergraduate teaching-oriented universities under the jurisdiction of the Ministry of Education (MOE), the Korean government conceived a dual system, with the Ministry of Science and Technology (MOST) also being permitted to found a research-oriented institution specializing in science and technology. This was accomplished with the establishment of the Korea Advanced Institute of Science and Technology (KAIST) in 1975; this institution subsequently played a pivotal role in generating a large number of high caliber scientists and engineers.

Another important effort to upgrade the R&D capability of universities throughout the country began in the late 1970s and 1980s. The government established the Korea Scientific and Engineering Foundation under MOST in 1977. In a complementary move, the government also established the Korea Research Foundation under MOE in 1981. These foundations funded basic research in universities. The government also enacted the Basic Research Promotion Law in 1989, explicitly targeting basic research as one of the nation's top technological priorities. As a result, as shown in Table 3, the size of industry-supported university research increased from W1.0 billion (US$2.1 million) in 1975 to W169.2 billion (US$ 240 million) in 1990, reflecting the hugely increased interest of industry in tapping sophisticated capabilities available in universities. The proportion of industry-funded university research also increased from 50 percent in 1975 to nearly 70 percent in 1990. Nevertheless, except for a few universities, university R&D remained rudimentary compared with the increasingly sophisticated needs of industry.

In addition to the formal university-industry relations, there existed informal knowledge transfer mechanisms such as private consulting by university professors. These moonlighting arrangements, which were quite extensive, were flexible and cost effective in helping firms to strengthen their technological capabilities. Private firms could identify university faculty members whose expertise was specifically relevant to their technological tasks and recruit them on a short term basis. This form

TABLE 3: SOURCES OF FUND FOR UNIVERSITY AND GRI RESEARCH

	1975	1980	1985	1990	1994
For universities					
(a) Government	0.9	10.1	30.8	74.7	170.4
(b) Industry	1.0	15.6	87.7	169.2	437.9
(c) Others	0.1	0.1	0.3	0.4	0.6
(d) Total	2.0	25.9	118.8	244.3	608.9
(b)/(d)	50.0	61.4	73.8	69.5	71.9
For GRIs					
(e) Government	37.9	94.0	271.7	537.4	981.4
(f) Industry	5.6	7.9	93.9	193.3	557.7
(g) Others	0.2	2.6	1.7	0.3	1.5
(h) Total	43.7	104.5	367.2	731.0	1,540.6
(f)/(h)	12.8	7.5	25.6	26.4	36.2
(a)/(e)	2.4	10.7	11.3	13.9	17.4
(b)/(f)	17.9	197.5	97.3	87.5	78.5
(d)/(h)	4.6	24.8	32.4	33.4	39.5

Note: Errors in total are due to round-up.
Source: Korea Industrial Technology Association, *Major Indicators of Industrial Technology*, Seoul: KITA, various years.

of collaboration may be symptomatic of Korea's position as a catching-up country, where technological tasks are not yet at the frontier (Kim 1993).

Another important feature of university-industry relations was the emergence of technology-based venture firms, as in advanced countries (Brett *et al.* 1991; Roberts 1991; Jones-Evans and Kirby 1995). KAIST and several leading universities have served as incubating institutions for such venture firms. One example is Medison, one of many technology-based small firms spun off from KAIST, and now one of the most successful new venture firms in Korea. The founder, who had a PhD in electronics engineering, and his four cofounders were graduate students in a laboratory that undertook research at KAIST on ultrasonic scanner technology. Medison is now one of the most dynamic ultrasonic scanner producers in the world, accounting for 25 percent of the global market in portable models. The firm has expanded its production and marketing activities worldwide and diversified into related businesses such as bioenergy medical equipment and medical information systems (Kim 1997a).

In parallel with the aforementioned investment to modernize university R&D activities, the government established several specialized GRIs, focusing on such sectors as chemicals, machinery, electronics, ocean science, standardization, nuclear energy, biotechnology, system engineering, aerospace, to serve the growing needs of the private sector. GRIs have continued to play an important role in strengthening the bargaining power of local enterprises in acquiring increasingly sophisticated foreign technologies. Although Korean firms have acquired successively more sophisticated capabilities, substituting their own domestic technologies for foreign technologies, they have had to rely continuously on imports for new technologies in almost all industries. For instance, when Corning Glass refused to transfer optical fiber production technology to Korea in 1977, two large copper cable producers in Korea entered a joint R&D project with the relevant GRI. After 7 years of R&D, the locally developed optical cable was tested successfully on a 35-km route in 1983. Although this local effort eventually grounded to a halt due mainly to slow progress in R&D, it nonetheless helped local firms gain bargaining power in acquiring foreign technology on favorable terms. Four firms entered licensing agreements with multinational enterprises in 1984 (Kim 1993).

In addition, the government introduced two major national R&D projects: the Industrial Generic Technology Development Project (IGTDP) and the National R&D Project (NRP). The IGTDP has concentrated mainly on solving current problems in "existing" technology areas with high economic externalities (i.e. "spillover effects"). The Ministry of Trade, Industry, and Energy (MOTIE) undertakes a survey each year to identify urgent R&D projects in industrial firms and offers financial supports to GRIs and university laboratories to undertake the projects jointly with private firms. Most of them are related to import-substitution of Japanese components in the electronics and machinery industries. In a 1989 survey, for instance, 174 technologies were identified, 146 of which were designated as projects to be funded. For IGTDPs, the government earmarked W11.5 billion (US$17.2 million) in 1989.

In contrast, NRP projects focus primarily on future problems in "new" (to Korea) technology areas with a high risk of failure or with high economic externalities, thus warranting public support. MOST identified several target areas, such as localization of machinery parts and components, new materials development, semiconductor design, super-mini computer development, energy conservation technology, localization of nuclear energy fuel, new chemical development, biotechnology development, and basic research in universities. The government's total investment in NRPs increased significantly from W13.3 billion (US$17.7 million) in 1982 to W98.8 billion ($123.5 million) in 1993.

DYNAMIC SURGE OF R&D IN KNOWLEDGE-INTENSIVE STAGE INDUSTRIALIZATION

In the period when technological upgrading to knowledge-intensive industries is proceeding in earnest, the relations between the different institutions of R&D activity change once again, under the impress of an internal dynamic of technological learning in the country concerned, and the external pressures of more and more intense competition at these technologically advanced levels. While the role of universities is enhanced, much of the product-oriented R&D is now accomplished in rapidly expanding R&D laboratories of private firms. This phase is now clearly established in the case of Korea.

In the wake of growing importance of innovation capability in sustaining Korea's international competitiveness in recent years, the private sector drastically stepped up its R&D efforts, from W2.37 trillion ($3.36 billion) in 1990 to W6.90 trillion ($8.95 billion) in 1995, maintaining its proportion to the nation's total R&D at 81 percent during the same period (see Table 2). This is one of the highest levels among both advanced and newly industrialized countries. This is significantly higher than the levels of 50.5 percent in the United States, 52.1 percent in Taiwan, 54.4 percent in Singapore, 21.7 percent in Greece, 40.3 percent in China and 77.4 percent in Japan. The Table also shows that the number of corporate R&D centers increased from 966 in 1990 to 2,270 in 1995, reflecting the seriousness attached to R&D by private firms in recent years.

Three examples show how Korea has now drawn abreast of the world technology frontier in certain sectors. In semiconductors, Samsung developed the 256 Mega dynamic random access memory (DRAM) chip ahead of Japan (Kim 1997b). Korean firms are neck-and-neck with Japanese firms in developing flat panel display technologies. Although a core patent was licensed from the US, Korea is one of the first countries that commercialized code division multiple access (CDMA) mobile telephone technology. It is private sector R&D that underpins these and similar achievements.

In this process, Korean firms have boosted their own capabilities through recruitment of highly trained Korean-American scientists and engineers. The US is populated with thousands of top-notch scientists and engineers who are Korean-Americans. Leading *chaebols* have lured away some of the best. Many of them left Korea more than a decade ago, earned PhDs in America's best universities, and rose through the ranks of such leading US concerns as IBM, Fairchild, Intel, and National Semiconductor. The well-financed Korean *chaebols* gave them challenging jobs and attractive compensation packages with considerable independence. Government statistics show that the number of scientists and engineers recruited by corporate R&D centers from abroad was 427 in 1992 alone. Some came back for short-term assignments, indicating that many Korean scientists and engineers abroad maintain close technical ties with Korean firms.

In addition to intensified in-house R&D, Korean firms began globalizing their R&D activities. LG Electronics, for instance, has developed a network of R&D laboratories in several locations, including Tokyo, Sunnyvale in California, Chicago, Germany, and Ireland. These outposts monitor technological change at the frontier, seek opportunities to develop strategic alliances with local firms, and develop the state-of-the-art products through advanced R&D. The LG Technology operation in Sunnyvale, for instance, plays a pivotal role in designing the latest personal computers, display terminals, and high resolution monitors, while the LG North American Laboratory in Chicago concentrates on HDTV, digital VCR, and telecommunication equipment. Samsung, Daewoo, and Hyundai Electronics have developed equally extensive R&D outposts. Samsung has R&D outposts in San Jose, Maryland, Boston, Tokyo, Osaka, Sendai in Japan, London, Frankfurt, and Moscow. Daewoo has two in France, one in the UK, and one in Russia. Hyundai has outposts in San Jose, Frankfurt, Singapore, and Taipei.

Korean firms are also globalizing R&D through mergers and acquisitions. Hyundai, for example, has been the most aggressive at acquiring equity stakes in foreign firms as a way to gain access to cutting edge technologies. For instance, Hyundai acquired full ownership of Axil Computer in Santa Clara, California for computer development;

it acquired significant stakes in Laserbyte Corp. in Sunnyvale, California to gain access to magneto optical disk drive technology, in Metaflow in La Jolla, California to develop SPARC compatible microprocessors, in Image Quest in San Jose, California to develop flat panel displays, and in Maxtor in San Jose, California to develop hard disk drives (Kim 1997a).

University research has also expanded significantly. Table 2 shows that R&D expenditure by universities almost tripled in five years from W244.3 billion in 1990 to W770.0 billion (US$999.5 million) in 1995. The number of university researchers also more than doubled from 21,332 to 44,683 during the same period. In addition, emulating the US experience, the government introduced in 1989 a scheme to establish Science Research Centers (SRCs) and Engineering Research Centers (ERCs) in the nation's leading universities. The number of SRCs and ERCs increased from 13 in 1990 to 35 in 1995. These centers receive research grants from the government for nine years.

The knowledge-intensive stage also saw the resurgence of strong university-industry linkages. Table 3 shows, for instance, that the total amount of university research support by industry increased significantly in four years from W169.2 billion in 1990 to W437.9 billion (US$567.9 million) in 1994. The number of joint research projects undertaken by SRCs and ERCs with industry also increased from 24 cases involving 34 firms in 1990 to 415 cases involving 338 firms in 1994 (KITA 1995), indicating a drastic increase in the scale and intensity of university-industry relations. In addition, several leading universities (i.e. Seoul National, Yonsei, and Korea) have developed science parks on the campus. Several *chaebol* firms are in the process of establishing joint-research laboratories in these universities.

In the government programs, GRIs continue to serve as the backbone of advanced R&D in Korea. This is well reflected in the fact that over 80 percent of national R&D investment goes to GRIs. Some of the GRIs have developed significant research results (such as 4M DRAM memory chips, electronics switching system, CDMA mobile telephone system) which were subsequently passed across to the private sector. Nevertheless, in the face of the rapid expansion of private R&D activities and increasing intensity of university R&D, reform of GRIs to redefine their roles has been discussed for sometime. Organizational inertia and the labor union of GRI members have made it difficult to implement these reforms.

In addition to the two national R&D programs mentioned above, the government introduced in 1992 the Highly Advanced National R&D (HAN) Project, also known as the G-7 Project, which is aimed at lifting Korea's technological capability to the level of G-7 countries by the year 2020. These three major national R&D projects (there are also three others) are designed to encourage GRIs to enter into consortia with the private sector. Universities also have access to these projects.

HAN, the most ambitious government program, has two parts: product technology development projects and fundamental technology development projects. The former includes new drugs and chemicals, broadband integrated services digital network (B-ISDN), next generation vehicle technology, and high definition television (HDTV). The latter projects cover such activities as ultra-large scale integrated circuits, advanced manufacturing systems, new materials for information, electronics, and energy industries, environmental technology, new functional biomaterials, alternative energy technology, and next generation nuclear reactors. Altogether, a total of US$5.7 billion will be invested jointly by the government, universities, and industries, about half of which will come from the private sector.

The sum of $1.3 billion has already been invested during the first three years, involving over 13,000 researchers and resulting in 2,542 patent applications and almost two thousand academic articles. Notable outcomes include quinolon-based antibiotics, liver disease treatment medication, and HDTV. The 256-Mega DRAM chip development was also one of the target technologies designated by the HAN, but private producers had already built enough technological capability to develop the chip on their own.

The most recent program introduced by the government to enhance Korea's basic capability is the "Creative Research Initiative Program", introduced in 1997. This program, albeit small in terms of investment size in the initial year, is designed to identify two dozen or so promising young scientists and engineers and provide them with sufficient research grants for nine years, subject to two interim reviews, to undertake concentrated advanced research so as to make breakthroughs in frontier technologies. Researchers from both universities and GRIs have shown enthusiastic responses.

Thus, the role of the R&D community has become increasingly important in the most recent phase of industrialization, as Korea entered relatively more technology-intensive industries. Universities play more important roles not only as a major source of well-trained scientists and engineers, but also as a major source of advanced research breakthroughs in frontier technologies. They also incubate technical entrepreneurs, who, with appropriate encouragement, spin off technology-based small ventures. While GRIs may have to undergo a drastic reform in the years to come, they are expected to play an increasingly important role in the area of public technologies such as nuclear energy, environment, health care, and marine science. They are also expected to provide effective technical services to small- and medium-sized enterprises (SMEs). In short, in the presence of demand from the private sector, the supply side of technological capability from the R&D community becomes effective, provided that the R&D community keeps reforming its role in providing effective services to the private sector with expertise, agility and flexibility.

CONCLUDING REMARKS

Korea has, indeed, been transformed from a subsistent agrarian economy into a newly industrializing one during the past three decades. Such phenomenal industrial development stemmed largely from the rapid acquisition of technological capability in the process from imitative "learning by doing" to innovative "learning by research" in the course of continuous market and technological change. The evolution of roles and dynamic interactions among the private sector, universities, and GRIs have made major contributions to the Korean industrialization. Such transformation offers many lessons for other developing countries, which are eager to see if Korean experience could be emulated in their countries. Seven such lessons are worth identifying.

First, Korean experience indicates that the first and most important implication for public policy is expanded investment in education even before launching an industrialization program. Compared with other developing countries that attained an equally rapid growth in elementary education, what was unique in Korea was the well-balanced expansion in all levels of education early enough to support its economic development. If anything, Korea went too far in this direction: the expansion of education exceeded the pace of economic development and created a short-term unemployment problem in Korea; the number of graduates in most fields exceeded demand. Consequently, unemployment among the educated was regarded as a serious social problem in the 1960s. But the formation of educated human resources laid an important foundation not only for the subsequent imitative reverse-engineering in the initial stage and advanced reserve-engineering in the intermediate stage, but also for rapid assimilation of imported foreign technologies.

Second, the most fundamental way to strengthen indigenous technological capability is to make definitive commitments to expanding tertiary educational programs and raising their quality. Building a significant number of first-rate educational institutions requires enormous financial and intellectual investment and requires a decade or more. This is one of the several major mistakes made by Korea. Korea expanded its tertiary educational program but failed to invest sufficiently to upgrade its quality. The Korean government belatedly recognized the mistake and has attempted to introduce a major reform in recent years, but it will take a decade or so before its effects can be seen.

In parallel with the expansion of tertiary educational programs, developing countries need to introduce research funding programs to strengthen basic research capability. Policymakers in developing countries often overlook the importance of basic research, which in their case is not designed to create new technologies to challenge advanced countries, but rather to provide a "window of opportunity" for them to catch-up with advanced countries. Baumol, Blackman, and Wolff (1991) also conclude that the quantity and quality of education in an economy is one of the major influences determining whether the economy is catching-up rapidly to narrow the gap with advanced countries. Developing countries achieved parity with advanced countries in terms of the percentage of children attending primary school. The provision of secondary and higher education explains differences in national wealth. Governments in developing countries should assume full responsibility for taking necessary measures to provide quality secondary and tertiary education to their populace.

Third, it is important to adopt a liberal policy with respect to the phenomenon of "brain drain" in developing countries. It makes sense to allow science and engineering graduates to migrate to advanced countries. Otherwise, many of them will not find suitable jobs and continue to advance their technical competence at home in the early stage of industrialization. Brain drain was also a serious problem for Korea through the 1960s. As of 1967, 96.7 percent of Korean scientists and 87.7 percent of Korean engineers educated abroad remained there, mainly in the US, compared with the corresponding world comparisons of 35 and 30.2 percent for all countries (Hentges 1975). They, however, became important sources of an overseas technical network and a high calibre manpower pool for Korea's subsequent development. These high-ranking scientists and engineers enabled Korea to grow as a contender in cutting-edge technologies in the knowledge-intensive stage.

For this purpose, governments in developing countries need to introduce, at the appropriate stage, a program to reverse brain drain to attract scientists and engineers residing abroad to return home. The Korean government launched an ambitious program to repatriate overseas Korean scientists and engineers living abroad in the 1960s and 1970s. Such a program also set a model for the private sector, which assertively recruited high-caliber scientists and engineers in the 1980s and 1990s. Taiwan also implemented a similar program to lure home overseas Chinese scientists and engineers.

Fourth, although S&T is not important for developing countries to acquire and assimilate mature foreign technologies in the initial stage of industrialization, it is imperative to invest in developing S&T infrastructure such as GRIs in the early stage. The Korean experience shows that it takes a decade or longer to develop an effective S&T infrastructure. In the early stage of industrialization, S&T infrastructure, particularly GRIs, suffer from poor linkages with industries.

The most important role of S&T infrastructure in the early stage, albeit unintended, was to generate experienced researchers, when the private sector faltered in R&D investment. Then when large firms began establishing corporate R&D centers to respond to market competition, these experienced researchers within the S&T infrastructure could be recruited and played a pivotal role in these private R&D centers.

Fifth, one of the most significant transformations in the Korean experience is the private sector's drastic shift from a minimal role in the nation's R&D in the initial stage to its dominant role in the knowledge-intensive stage. An implicit point is that such a transformation was possible primarily in response to intensifying competition in the market. Korean experience shows that the most effective source of competition is export policy. That is, it is imperative for developing countries to adopt export promotion policy if they want to expedite industrialization. This policy creates business opportunities and concurrently imposes crises for firms to undergo a "life or death" struggle in the competitive international market. The Korean experience shows that to survive in the crises, Korean firms had to accelerate learning by importing and rapidly assimilating production technology from abroad.

Consequently, firms in export-oriented industries were forced to learn more rapidly and grew faster than firms in import-substituting industries. Likewise, countries with export-oriented industrialization (EOI) grew faster than those with import-substituting industrialization (ISI). For instance, the average annual economic growth rate for EOI countries was 9.5 and 7.7 percent, respectively, for 1963–1973 and 1973–1985 periods, as compared to 4.1 and 2.5 percent for ISI countries. The real per capita income growth rate was 6.9 and 5.9 percent for the same periods for the former as compared to 1.6 and − 0.1 for the latter, as the latter group had a higher population growth rate. It can be argued that this was the principal reason why the EOI-oriented NICs in East Asia grew faster than their ISI-oriented counterparts in Latin America.

Sixth, it is also important for governments in developing countries to offer various incentive programs for the private sector to undertake R&D activities. These programs may include tax concessions, preferential financing, and the exemption of military service obligation for key R&D personnel. The first two enable firms to reduce the cost of setting-up R&D laboratories and undertaking R&D activities, while the last one helps firms to recruit high caliber young technical people who can commit to the R&D function.

Finally under the World Trade Organization (WTO) regime, it is not easy to regulate the flow of foreign technology or investment. Nevertheless, it is important for developing countries to recognize that in the early stage of industrialization, foreign licensing (FL) and foreign direct investment (FDI) are not so important for acquiring foreign technologies. The really significant means for technology acquisition is the procurement of turnkey plants (in the case of continuous process industries such as chemicals) and capital goods such as plant and equipment. The Korean experience was that FDI and foreign licensing was restricted in the early years, while technology transfer through other means such as capital goods imports was promoted in the early years. Such a policy, designed to maintain Korea's management independence from foreign multinationals, was effective in forcing Korean firms to take initiatives and give a central role to learning (i.e. acquiring, assimilating, and improving imported technologies) rather than relying entirely on foreign sources. Well-trained, hard working Koreans were motivated to maximize technological learning from readily available foreign goods and were equipped with sufficient knowledge to reverse-engineer them successfully.

In conclusion, despite evident differences in specific sociocultural, economic, and technological conditions in developing countries, there appear to be many lessons to learn from Korea, which has undergone a rapid transition from imitative "learning by doing" to innovative "learning by research" in R&D. The lessons involve learning from Korea's successes, while not repeating Korea's mistakes.

REFERENCES

Baumol, W.J., Blackman, S.A. and Wolff, E.N. 1991: *Productivity and American Leadership*. Cambridge, MA: The MIT Press.

Brett, A.M., Gibson, D.V. and Smilor, R.W. (eds) 1991: *University Spin-off Companies—Economic Development, Faculty Entrepreneurs and Technology Transfer*. Savage, MD: Rowman and Littlefield.

DIST 1994: *Australian Science and Innovation Resources Brief 1994*, Department of Industry, Science and Technology, Paper No. 21. Canberra: Australian Government Publishing Service.

Harbison, F.H. and Myers, C.S. 1964: *Education, Manpower, and Economic Growth*. New York: McGraw-Hill.

Hentges, H.A. 1975: *The Repatriation and Utilization of High-Level Manpower: A Case Study of the Korea Institute of Science and Technology*, unpublished PhD dissertation, John Hopkins University.

Jones-Evans, D. and Kirby, D.A. 1995: The formation and development of small technical professional service providers in the north-east England—a preliminary study, *Entrepreneurship and Regional Development*, 7: 21-40.

KDB (Korea Development Bank) 1994: *Korean Industry in the World, 1994*. Seoul: Korea Development Bank.

Kim, L. 1991: Pros and cons of international technology transfer: A developing country's view, in T. Agmon and M. A. von Glinow (eds) *Technology Transfer in International Business*. New York: Oxford University Press, 223-239.

Kim, L. 1993: National system of industrial innovation: Dynamics of capability building in Korea, in R. Nelson (ed) *National Innovation Systems: A Comparative Analysis*. New York: Oxford University Press, 357-383.

Kim, L. 1997a: *Imitation to Innovation: The Dynamics of Korea's Technological Learning*, Boston: Harvard Business School Press.

Kim, L. 1997b: The dynamics of Samsung's technological learning in semiconductors, *California Management Review*, 39 (3): 86-100.

Kim, L. and Kim, Y. 1985: Innovation in a newly industrializing country: A multiple discriminant analysis, *Management Science*, 31 (3): 312-322.

Kim, L. and Lee, H. 1987: Patterns of technological change in a rapidly developing country: A synthesis, *Technovation*, 6 (4): 261-276.

KIST 1994: *KIST 25 Nyunsa (25 Years History of KIST)*. Seoul: Korea Institute of Science and Technology.

KITA (various years), *Major Indicators of Industrial Technology*. Seoul: Korea Industrial Technology Association.

KITA 1995: *Sanup Gisul Baekseo, 1995* (Industrial Technology White Paper, 1995). Seoul: Korea Industrial Technology Association.

Magaziner, I.C. and Patinkin, M. 1989: Fast Heat: How Korea won the microwave war, *Harvard Business Review*, January-February 83-93.

Pack, H. and Westphal, L.E. 1986: Industrial strategy and technological change, *Journal of Development Economics*, 4: 205-237.

Roberts, E.B. 1991: *Entrepreneurs in High Technology: Lessons from MIT and Beyond*. New York: Oxford University Press.

Utterback, J.M. 1975: The role of applied research institutes in the transfer of technology in Latin America, *World Development*, 3 (9): 665-673.

[12]

Absorptive Capacity and Industrial Growth: A Conceptual Framework and Korea's Experience

Linsu Kim

INTRODUCTION

Korea's rapid industrialization may be attributed to many economic, social and technical factors. The most important of all may be absorptive capability, which is a combined outcome of various economic, social and technical inputs. Absorptive capability enables one to assimilate, use, adapt and change existing technologies. It also enables one to create new technologies and to develop new products and processes in response to a changing economic environment. Absorptive capability also includes capacity to acquire, adapt and internalize managerial know-how. This latter element of absorptive capability, however, is not explicitly covered in this chapter.

The economic development of a country is, in fact, the sum of the production of organizations and individuals. Since organizations, particularly firms, play a major role in industrialization, this chapter deals primarily with organizational absorptive capability and its relationship with both external and internal elements.

This chapter first presents, albeit in brief, a conceptual framework that depicts the process associated with the acquisition of absorptive capability at the organization level. Then, using the framework as an analytical tool, it analyses Korea's past progress and its current problems/future prospects in industrialization. Finally, it attempts to draw implications for other developing countries.

I ABSORPTIVE CAPABILITY: A CONCEPTUAL FRAMEWORK

Absorptive capability not only requires learning capabilities but also develops problem-solving skills. Learning capabilities involve development of the capacity to assimilate existing knowledge, while problem-solving skills represent a capacity to create new knowledge.

Absorptive capability has two important elements: prior knowledge related to the task and intensity of effort. First, prior knowledge is an essential prerequisite to absorptive capability, as learning is cumulative. Accumulated prior knowledge increases both the ability to make sense of, to assimilate and to use new knowledge. Relevant prior knowledge includes basic skills in the case of less developed countries (LDCs). But it also includes knowledge of the most recent scientific and technological knowledge in the case of industrially advanced countries (IACs). For instance, relevant prior knowledge required to assimilate technologies associated with labor-intensive, simple products and processes (e.g., plywood-manufacturing) is drastically different from that required to assimilate technologies associated with knowledge-intensive complex products and processes (e.g., robotics design and manufacturing). Thus, prior knowledge should be assessed in its relation to the task difficulty involved.

Second, the intensity of effort is critical in accumulating absorptive capability, as it is insufficient merely to expose firms to the relevant knowledge without exerted efforts to internalize such knowledge. Learning how to solve problems is usually built up over many practice trials on related problems (Harlow, 1959). Thus, it requires considerable time and effort on early problems before moving on to more complex problems in order to accumulate absorptive capability. Although Max Weber discussed the importance of prevalent attitudes such as willingness to work hard, neoclassical economics has shunned it (Stiglitz, 1991b).

These two elements of organizational absorptive capacity are analogous to Lewin's model – Performance = f (ability, motivation) – of individual performance in the organization. One has to have an array of ability or knowledge relevant to given tasks and at the same time motivation to exert effort in order to produce excellent performance. Likewise, an organization's absorptive capacity to perform can be said to have similar elements.

The process of accumulating knowledge in an organization, often referred to as organizational learning, is not a simple sum of individual learning, although individuals are the agents through whom the learning takes place. Rather, it is knowledge that is distributed across the organization, is communicable among members, has consensual validity, and is integrated into the working procedures and administrative structures of the organization (Duncan and Weiss, 1978). The process of knowledge accumulation through organizational learning is influenced by a broader set of social, political and structural variables in the organization

(Shrivastava, 1983). Hedberg (1981) and two recent issues 2(1)(1991) and 3(1)(1992) of *Organization Science* carry a number of excellent articles on the concept of organizational learning and relations between individual learning and organizational learning.

Likewise, the intensity of effort at the organizational level is not a simple sum of individual motivation in the organization, although effort is exerted by individual members in the organization. Rather, it is influenced by a complex process of group dynamics and organizational culture, which are in turn determined to a large extent by organizational size, membership composition, members' norms, organizational tasks, leadership style, group cohesiveness, reward systems, and administrative structure. Napier and Gershenfeld (1985) provide a useful theoretical review on how the group affects individual behavior and Schein (1985) provides an excellent review on how individual behavior in the organization is shaped by organizational culture and how organizational culture may be changed.

These two variables, prior knowledge and the intensity of effort in the organization, constitute, as presented in Figure 13.1, a 2x2 matrix that indicates the level of absorptive capability. When both prior knowledge related to task difficulty and the intensity of effort are high (Cell 1), absorptive capability is high. On the contrary, when both elements are low (Cell 4), absorptive capability is low. Organizations with high prior knowledge related to task difficulty and low intensity of effort (Cell 2) will gradually lose absorptive capability, as prior knowledge will become obsolete as task-related technologies move along its trajectory. Such organizations will gradually lose absorptive capability, moving down to Cell 4. In contrast, organizations with low prior knowledge related to task

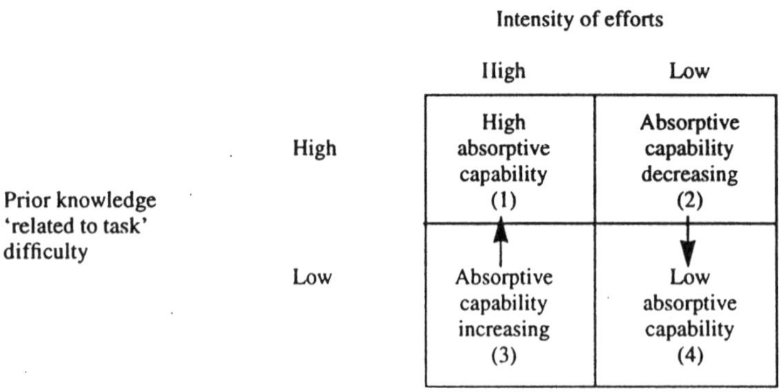

Figure 13.1 Two elements of absorptive capability

difficulty and high intensity of effort (Cell 3) will more rapidly be able to acquire absorptive capability, as repeated efforts to learn and to solve problems elevate the level of relevant prior knowledge, moving gradually to Cell 1.

Absorptive capability is acquired and accumulated mainly through two cycles or interactive mechanisms: 'boundary spanning cycle' and 'internal cycle'. 'Boundary spanning cycle' raises firms' absorptive capability through their interactions with the environment, while 'internal cycle' gives rise to their absorptive capability through production and R&D within the organizations. Further, these two cycles intersect and reinforce each other. Effective 'boundary spanning cycle' facilitates 'internal cycle', while effective 'internal cycle' promotes 'boundary spanning cycle', both through accumulated absorptive capability.

'Boundary spanning cycle', as presented in Figure 13.2, shows how existing absorptive capability enables firms to exploit external information and how external information, in turn, gives rise to firms' absorptive capability. First, absorptive capability includes not only substantive, technical knowledge, but also awareness of where useful complementary expertise is available outside the organization (Cohen, W. and Levinthal, 1990). Firms that develop a broad and active network of external relationships strengthen individuals' awareness of others' capabilities and knowledge and the firm's absorptive capability. Kim (1988a) shows that innovative firms in Korea characteristically develop more extensive external relations with foreign technology suppliers, buyers, local universities, and public research institutes than less innovative firms. Von Hippel (1988) also shows the importance of close relationships with both suppliers and buyers in the United States. Absorptive capability also enables firms to recognize the value of new external information, assimilate it, and apply it to commercial ends. It also makes the firm more proactive with a high sensitivity to emerging technological opportunities, aggressively seeking out new opportunities to exploit and develop its technological capability. Absorptive capability also enables the firm to predict more accurately the nature and commercial potential of technological progress (Cohen, W. and Levinthal, 1990).

Second, external information, as transferred to firms, strengthens their absorptive capability by raising the level of prior knowledge, which is an essential component of absorptive capability. This process is often called 'technology transfer'. Human-embodied external information (e.g., newly recruited scientists and engineers who have had extensive training and experience in advanced countries) gives significant rise to prior knowledge of firms in developing countries (Kim, 1991). Machine-embodied

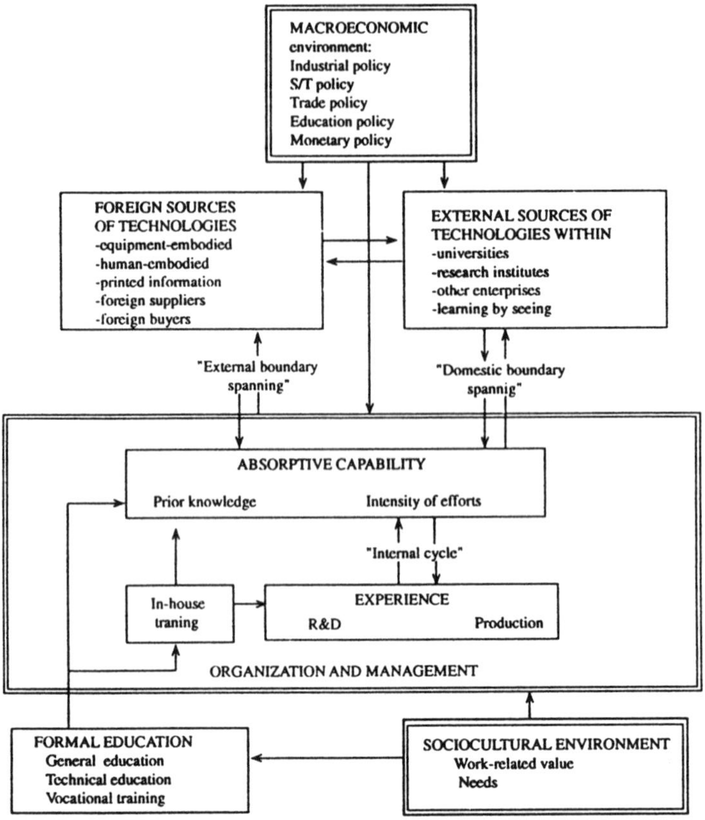

Figure 13.2 Process of absorptive capability accumulation

information is a critical mode of technology transfer for firms to undertake imitative reverse engineering in developing countries. Printed information such as sales catalogues, blueprints, technical specifications and trade journals serves as an important source of new knowledge for firms in newly industrializing economies (NIEs) (Kim, and Kim, 1985). Technology and equipment suppliers provide critical information, while OEM (original equipment maker) buyers often transfer critical knowledge to producers to ensure that products meet technical specifications.

Absorptive capability may also be raised through domestic boundary spanning from sources within the economy. Local universities, public research institutes and other enterprises within the economy could be important sources of scientific and technical information new to firms.

Joint research projects with universities and research institutes could give a significant rise to the technological capability of a firm. The observation of advanced equipment in use by other enterprises or in exhibition at trade fairs is a major source of imitative reverse engineering or 'learning by seeing' (Kim, L. and Kim, 1985). In short, absorptive capability enhances the ability to exploit external information, while transfer of external information, in turn, strengthens firms' stock of prior knowledge, raising absorptive capability.

In addition, industrial, trade, science/technology, and monetary policies set macroeconomic environments in which firms have to operate, thus affecting directly and indirectly the process of absorptive capability accumulation through boundary spanning cycles.

The internal cycle, as shown in Figure 13.2, presents the way in which absorptive capability strengthens production and R&D trials, in turn, raise prior knowledge of the individuals involved and of their firms, thereby raising absorptive capability. First, the richness of previous knowledge increases the ability to assimilate external information, which can be translated into production for firms in developing countries, particularly in undertaking imitative reverse engineering. Prior possession of relevant knowledge and skills raises creativity in R&D, permitting the sorts of associations and linkages that may have never been considered before (Simon, 1985).

Second, absorptive capability may be developed as a by product of production operations through learning by doing. This is particularly important for firms in developing countries. The more organizations become practised, the more they become capable at activities in which they are already engaged. In addition, production experience enables firms to recognize and exploit new information relevant to particular production methods to improve their own processes. Absorptive capability is also generated as a by product of a firm's R&D, particularly when knowledge is less explicit and codified (Nelson and Winter, 1982) and more difficult to assimilate in a technologically progressive environment. The more difficult learning is, the more prior knowledge has to have been accumulated via R&D for effective learning to occur. The more difficult the learning environment, the higher marginal effect of R&D on absorptive capability (Cohen, W. and Levinthal, 1989). R&D activities will enable firms to be aware of the significance of new external signals and exploit available information more effectively. In short, absorptive capability promotes R&D and production activities, while production experience and R&D trials, in turn, give rise to absorptive capability.

The kind, size and state of the formal education system effect the supply of human resources that embody the basic capacity and skills firms have to deploy in both boundary spanning and internal cycles. Sociocultural environment forms the norms, values and needs of organizational members, who are involved in absorptive capability accumulation. Lastly, the way the organization is structured and managed sets the tone within the organization that significantly affects both boundary spanning and internal cycles.

This conceptual framework will be used as a tool to analyse the past progress and future problems of Korea's industrialization.

II ANALYSIS OF PAST PROGRESS

Why was Korea so successful in its rapid industrialization in the 1960s and 1970s? Various factors that appear in the conceptual framework presented above may be attributed to such a growth.

First, ambitious investment in formal education by both government and parents during the 1950s and 1960s made possible the formation of relevant prior knowledge for industrialization in the early period. At the end of the Japanese colonial rule, only 2 per cent of the Korean population over 14 years of age had completed secondary school and the illiteracy rate stood at 78 per cent. But the government assigned high priority to education, raising the share of education in the total government budget, for instance, from 2.5 per cent in 1951 to over 17 per cent in 1966. Government expenditures, however, accounted for only one-third of the total expenditures in education, the remainder being borne by the private sector and parents, reflecting the high commitment to education within Korean society. As a result, enrollment in the various levels of the formal education system has increased rapidly since 1953. In the case of elementary school, enrollment has increased by over five times. Even faster growth is seen in the secondary and tertiary education levels.

Several other developing countries also attained as rapid a growth in elementary education as did Korea. But what was unique in Korea was the well-balanced expansion in all levels of education early enough to support its economic development. Harbison and Myers (1964) show that given GNP per capita from a regression line of 73 developing countries, Korea's level of educational achievement was furthest beyond the predicted level based on its economic development. That is, with a per capita income of $90, Korea's educational achievement stood fairly close to the normal pattern of human resource development for a country with a mean per capita GNP of $200. Korea's level of human resource development, when its per capita GNP reached $107, was equivalent to that of countries with a GNP per capita of $380. Overseas training and observation were also characteristic of human resource development in Korea. Overseas training was started back in the foreign aid period of the 1950s, when a high proportion of senior personnel in government, business and academia were exposed to foreign training, mainly in the US under economic assistance programs (Mason et al., 1980). Prior knowledge gained through formal education was so adequate for assimilating relevant machine-embodied or printed information that the need for additional in-house corporate training was minimal. Such an expanded educational system prepared Koreans to assimilate techniques and technology related to the erection of new plans and the operation of new machinery.

Second, the sociocultural traits. Korean's entrepreneurial and hardworking nature must have strongly affected the intensity of efforts within any given organization. Korea's land area is scarcely bigger than New Jersey or Hungary, crossed by mountains with a relatively small portion of arable land available for more than 40 million people. The cramped conditions and severely cold winters appear to have forced Koreans to work long and hard whenever possible in order to survive in an unfavorable environment. Besides, the older generation is motivated by the memory of deprivation and hard times under Japanese occupation and during the Korean War. A sense of *geug-il* or 'beat Japan' slogan to settle old scores and national economic competition with North Korea is also a major force motivating Koreans (Porter, 1990). Consequently, both policy-makers and managers had entrepreneurial 'can-do' spirit in the face of a series of economic and managerial crises. This spirit was, for instance, well demonstrated in two oil crises and Korea's challenge to new technologies. In addition, the average working hours per week in manufacturing as late as in 1985 were 53.8 hours in Korea compared to 33.1 to 42.9 in OECD countries including Japan, 44 to 48.0 hours in other Asian NICs, and 46.0 hours in Mexico (Kim, 1988a).

Third, industrial and trade policies had strongly affected the acquisition of relevant prior knowledge and the intensity of efforts to accumulate absorptive capability. While the Korean government protected local markets very carefully to foster infant-industry growth, it set forth exports as something of a life and death struggle in order to achieve economic growth goals with the small domestic market. Thus, the government pushed and pulled firms with threats and promises. That the Korean economy was primarily export-oriented appears to have affected the fast acquisition of absorptive capability in at least two ways. First, lump-sum investments for capacity beyond local market size to bring about the economy of scale forced local businesses to intensify the acquisition of absorptive capability in order to improve international competitiveness and in turn to maximize capacity utilization. Second, as producers entered the international market, the keen international competition forced them to intensify technological efforts significantly, mainly in learning by doing, reverse engineering and managerial/marketing efforts, in order to remain competitive in both quality and price.

Fourth, foreign technology transfer had been an important source of new knowledge for Korean firms. And public policy and a firm's cost-saving strategy shaped the structure of technology transfer to Korean firms. That is, Korea's policies concerning direct foreign investment (DFI) and foreign licensing (FL) were quite restrictive in the early years of industrialization. Consequently, the size of DFI and its proportion to total external borrowing were significantly lower in Korea than in other NICs. For example, Korea's stock of DFI in 1983 was only 7 percent of the size of that in Brazil, 23 percent of that in Singapore, and less than a half the size of that in Taiwan and Hong Kong. The proportion of DFI of total external borrowing was only 6.1 percent in Korea compared with 91.9 per cent in Singapore, 45 percent in Taiwan and 21.8 percent in Brazil (KEB, 1987). The comparative figures reflect Korea's explicit policy of promoting its 'independence' from multi-nationals in management control.

In contrast, Korea promoted technology transfer through the procurement of foreign capital goods. The rapid growth of the Korean economy has called for commensurate growth in investment for production facilities. Government policy had, however, been biased in favor of the importation of foreign capital goods and against the development of a rudimentary local capital goods industry as a way to strengthen international competitiveness of the capital goods user industries. Such a policy led to massive imports of foreign capital goods, which became a major source of relevant knowledge and learning through reverse engineering by Korean firms (Kim, L. and Kim, 1985). Of the four categories of technology transfer listed in Table 13.1, capital goods imports far surpass other means of technology transfer in terms of value. Capital goods imports were worth 23 times the value of DFI and 35 times the value of FL. The total value of capital goods imports is 14 times that of the other two categories combined, suggesting that Korea may have acquired more technology from advanced countries through the importation of capital goods than through any other means. Among NICs, the proportion of capital goods imports to total technology transfer was the

Table 13.1 Foreign technology transfer to Korea unit: US$ million

Source	*1962–66*	*1967–71*	*1972–76*	*1977–81*	*1982–86*	*1987–91*	*Total*
1 Direct foreign investment (DFI)							
Japan	8.3	89.7	627.1	300.9	875.2	2 113.6	4 014.8
USA	25.0	95.3	135.0	235.7	581.6	1 482.1	2 554.7
All others	12.1	33.6	117.3	184.0	309.7	2 036.1	2 694.8
Total	45.4	218.6	879.4	720.6	1 766.5	5 631.8	9 264.3
2 Foreign licensing (FL)							
Japan	–	5.0	58.7	139.8	323.7	1 483.9	1 911.1
USA	0.6	7.8	21.3	159.2	602.7	2 121.9	2 913.5
All others	0.2	3.5	16.6	152.4	258.5	853.6	1 284.7
Total	0.8	16.3	96.5	451.4	1 184.9	4 359.4	6 109.3
3 Technical consultancy							
Japan	–	12.	7.7	20.8	89.2	217.6	347.4
USA	–	3.1	6.0	16.7	159.1	619.8	804.7
All others	–	1.6	4.8	17.2	84.0	413.5	521.1
Total	–	16.8	18.5	54.7	332.3	1 250.9	1 673.2
4 Capital goods imports							
Japan	148	1 292	4 423	14 269	20 986	54 643	95 761
USA	75	472	1 973	6 219	12 394	33 099	54 232
All others	93	777	2 445	7 490	17 205	33 197	61 207
Total	316	2 541	8 841	27 978	50 585	120 939	211 200

SOURCE Korea Industrial Technology Association for DFI and FL; Korea Engineering Services Association for technical consultancy data; and Korean Society for Advancement of Machinery Industry for capital goods data.

highest in Korea (Westphal, Kim and Dahlman, 1985). In short, Korea restricted DFI but instead promoted technology transfer through other means such as capital goods imports. Such a policy, designed to maintain Korea's independence from foreign multi-nationals, has been relatively effective in acquiring technological capability in the mature industries, since Korea's well-trained human resources and their entrepreneurship enabled the country to learn quickly form foreign capital goods.

Fifth, both buyers and suppliers played an important role in enabling Korean firms to acquire absorptive capability. Informal technical assistance offered by foreign OEM (original equipment manufacture) buyers to ensure that Korean-made products met their technical specifications provided invaluable help to Korean firms in acquiring necessary relevant knowledge (Kim, 1991). Local buyers of capital goods provided useful information for custom-designed machinery. Technology and capital goods suppliers provided relevant knowledge to ensure successful implementation of their supplies (Kim, L. and Kim, 1985).

Sixth, the microeconomic behavior of Korean firms in acquiring absorptive capability varies significantly by industry. Kim and Lee (1987) provide a detailed discussion of how small batch industries differ from mass production and process-type industries in acquiring capability. A commonality across industries, however, is that given a large-size investment and lack of experience and capability, large firms have resorted heavily to foreign sources in order to ensure quick erection and smooth start-up of new plants. In contrast, small firms underwent a long process of imitative reverse engineering of foreign machinery, as they lacked financial resources and organizational capability to identify and negotiate with foreign technology suppliers. Nevertheless, both large and small firms deployed deliberate, aggressive technological strategies to acquire their own technological capabilities through efforts exerted to assimilate foreign technologies from the very outset. Korean firms have acquired technological capabilities in piecemeal fashion, as increasingly sophisticated capabilities have been acquired and put into practice. The process of acquisition has clearly been one of purposive efforts involving a succession of incremental steps, with production capabilities being developed somewhat in advance of engineering and innovation capabilities. The selectivity of import substitution for the elements of technology has meant continued reliance on imports for at least some elements in almost all industries, but the pattern of imports has continually shifted as local capabilities have replaced foreign ones (Westphal, Kim and Dahlman, 1985).

Seventh, firms' strategy in resource allocation in acquiring absorptive capability also worked well. Since R&D in the formal sense of the term was not important for Korea during this stage of imitating mature technologies, firms reversed the sequence of R,D&E: they started with engineering (E) for products and processes imported from abroad, and then progressively evolved into the position of undertaking substantial development (D). But research (R) was not relevant to Korea's industrialization through the 1970s. Several studies conducted in other countries also provide similar evidence. Studies of Japanese industrial history show that its industries went though a similar pattern (Shishido, 1972; Ozawa, 1974; Yakushiji, 1977). The United States also reversed the sequence of research, development and engineering; that is, it started with practical technologies imported from Britain, and then slowly evolved into the present position of being able to afford substantial research (NAS, 1973). Brazilian and Argentinian firms began manufacturing with assembly and packing and eventually extended into more intricate operations. Engineering was the initiating portion of the RD&E spectrum leading gradually to more advanced development and research efforts (Baranson, 1973; Dahlman and Frischtak, 1993; Katz and Bercovich, 1993).

Eighth, organizational structure and managerial style had also affected the acquisition of absorptive capacity. Although some 300,000 Koreans, mostly with menial jobs, had been involved in mining and manufacturing when Korea gained its independence in 1945, only a limited number of them had managerial experience. The Korean War in 1950 brought in the US military forces, which helped organize the Korean military. The military became a major source of organizational and managerial experience for most Korean males. Then in 1961, military leaders took over the government and introduced military bureaucracy into the Korean government system. And both military and government produced a large number of experienced managers for large enterprises, and the government intentionally helped the formation of and fostered the growth of large firms, chaebols, for the economy of scale. Influenced by the military, most firms were hierarchical, centrally controlled but relatively less formalized. Most male organizational members (females served only menial or assistant roles) were accustomed to such a climate through their experience in compulsory military service. Such organizations were quite compatible with and efficient in the acquisition of absorptive capability through learning by doing and in production (Kim and Utterback, 1983). Kim (1976) called such organizations 'implementation viable'. Nevertheless, unlike highly formalized organizations, they are adaptable to changes once a decision is made at the top.

Ninth, Korea's rapid growth had also contributed to the fast acquisition of absorptive capability. The faster the growth, the more the opportunities for learning by doing and the more the need to learn from external sources, with both giving rise to absorptive capability.

Tenth, another important feature of the Korean economy was the rapid diffusion of absorptive capability from existing organizations to new entrants through the mobility of experienced technical and managerial personnel. Kim (1980a) reports that about 70 percent of consumer electronics producers entered production through poaching experienced personnel. He also reports that engineers experienced at state-owned chemical and machinery enterprises in the early 1950s have spun off and played a pivotal role in emerging private firms (Kim, 1991).

In conclusion, the continuous inflow of foreign technology (through formal and informal mechanisms) to learn from, the formation of highly trained human resources who were willing to learn, entrepreneurs' 'can-do' spirit exercised freely under restricted equity participation of multinationals, and the government's orchestral role in directing firms and selectively allocating resources to them to achieve ambitious growth objectives under the pressure of the competitive international market have worked together for Korea to grow rapidly through the 1970s. Chaebol, particularly, played an important role in this process, as they were in the most advantageous position for drawing on the best quality human resources, in identifying, negotiating and financing foreign technology transfer, in obtaining business licences and preferential financing from the government, in applying experience gained in one field of business to another, and in risk-taking in new businesses with a cushion provided by existing businesses. As a result, by the end of the 1970s, Korea had the largest textile plant, the largest plywood plant, the largest shipyard, the largest cement plant and the largest heavy machinery plant in the world.

III ANALYSIS OF CURRENT PROBLEMS AND FUTURE PROSPECTS

Can Korea sustain its growth in the future? What are reasons for some of the current problems? What should Korea do for its future development? Using the conceptual framework presented above, this section attempts to answer these questions by analyzing the Korean economy,

First, in contrast to overinvestment for education in the 1950s and 1960s, one of the major mistakes made by the Korean government in national development is its underinvestment in higher educational institutions in the subsequent periods. The number of universities has increased significantly from 69 in 1966 to 100 in 1985. The number of university students has increased from 131,354 to 931,884, with the proportion of high school graduates who go on to universities growing from 28.2 percent to 38.3 percent during the same period. The environment for quality education and research has, however, deteriorated over this period. For example, the student–professor ratio has retrogressed from 22.6 in 1966 to 35.8 in 1985, making all universities primarily undergraduate teaching-oriented rather than graduate research-oriented. Thus, the 'publish or perish' principle is not applied. As a result, Korea has failed to develop the stock of highly trained scientists and engineers who would be necessary in order for Korea to sustain its international competitiveness in the 1990s. Compared to education at other levels, graduate education in science and engineering has been much neglected in terms of quantity and quality. The government's recent move to correct this discrepancy has resulted in across-the-board expansion of graduate programs in all universities in an attempt to meet future needs at least in quantity. It will take a great deal of time and effort, however, before such a move can produce quality output. In short, formal educational institutions in Korea are not sufficiently able to train and supply the human resources with relevant prior knowledge needed for creative imitation and innovation in the future.

Second, the traditional Korean traits of entrepreneurship, achievement orientation, and hard work in learning and at work have significantly deteriorated in the past decade mainly for three reasons. One, the new generation brought up in affluence is less willing to work hard compared to the older generation. Two, the recent move to democratize the economy and society triggered disorderly, violent and in many cases unlawful labor movements, resulting in organization anarchy, lowered morale, shortened work hours, work stoppage, lowered product quality and drastic wage hikes. Consequently, work attitude has degenerated drastically from 'work hard and save' to 'work less and enjoy'. With high aspirations for democracy and better job security by militant labor unions, workers refuse to work under a military-like, centrally controlled hierarchy. Three, such a drastic change in organizational climate has noticeably disheartened owners and managers, making them more reluctant to invest in uncertain projects than before. Such a drastic change in social/organizational climate has slowed down the pace of acquiring new relevant knowledge and weakened efforts to learn. Public policy-makers and corporate managers have the new task of motivating and managing the 'new generation'.

Third, the macroeconomic environment for Korea has changed significantly in the 1980s for several reasons. One, the world economy has generally slowed down in the 1980s, particularly affecting outward-looking economies like Korea. Two, in the wake of rising trade imbalances, North America and Europe have moved towards protectionist policies, making it increasingly difficult for Korea to sustain export growth in industries that have led Korea's export-oriented strategy in the past. Three, the United States has put pressure on Korea to open up its domestic market, making it unavoidably necessary for Korean firms to compete against multi-national enterprises in both international and domestic markets. Four, Korea has lost its competitiveness in low-wage-based labor-intensive industries, as its real wage has risen at the average annual growth rate of 5.8 percent in the 1960s and 7.5 percent in the 1970s. It also rose a cumulative 62.5 percent for three years since 1987. When adjusted in US dollar terms, the rise amounts to 91.1 percent, which is 2.8 times that of productivity (*Business Korea*, February 1990). Concomitantly, other developing countries (e.g., China, Indonesia, Malaysia and Thailand) with much lower wage rates are rapidly catching up with Korea in these labor-intensive industries. Five, Korea was forced to change its copyright and patent laws, pre-empting the imitative reverse engineering of foreign products. Such an unfavorable environment makes it imperative for Korea to enhance its own technological capabilities in order to innovate independently and to strengthen its bargaining power against foreign technology suppliers.

In response to such external forces, the government made a major policy shift. New policies include, among others, (i) economic democratization to mitigate monopolistic abuses by chaebol; (ii) open market policy, raising import liberalization ratio from 51 pe cent in 1973 to 95.2 in 1988 and lowering the average tariff rate from 26 percent in 1984 to 16 percent in 1988; and (iii) liberalization of foreign direct investment and licensing. The new macroeconomic environment has drastically reduced the ease of learning (or increased task difficulty) for Korean firms.

The economic problems Korea faces today are so complex that they defy simple analysis. However, the most critical problem simply put appears to be the negative gap between absorptive capability and task difficulty, in contrast to the positive gap between the two variables in the 1960s and 1970s that had been a major source of sustaining Korea's competitiveness. This is depicted in Figure 13.3. While a rapid change in Korea's macroeconomic environment in the 1980s has inevitably pushed up task difficulty for Korean firms, the unhealthy state of formal education and difficulty in learning has resulted in a significant decline in prior knowledge related to task difficulty. Coupled with the deterioration of job commitments stemming from militant labor unions and an affluent new generation, lowered relevant prior knowledge appears to have slowed down the pace of acquisition of absorptive capability in organizations. What concerns many is that the acquisition of absorptive capability is cumulative in nature and cannot be built overnight. Fourth, relevant external information, either machine-embodied or printed, is less explicit and codified and more difficult to learn than before. And firms in industrially advanced countries, particularly Japan, are increasingly reluctant to transfer technology to Korea. As a result, major channels for acquiring external information abroad have changed from machine-embodied to 'antennas' (technological outposts) and recruitment. In the

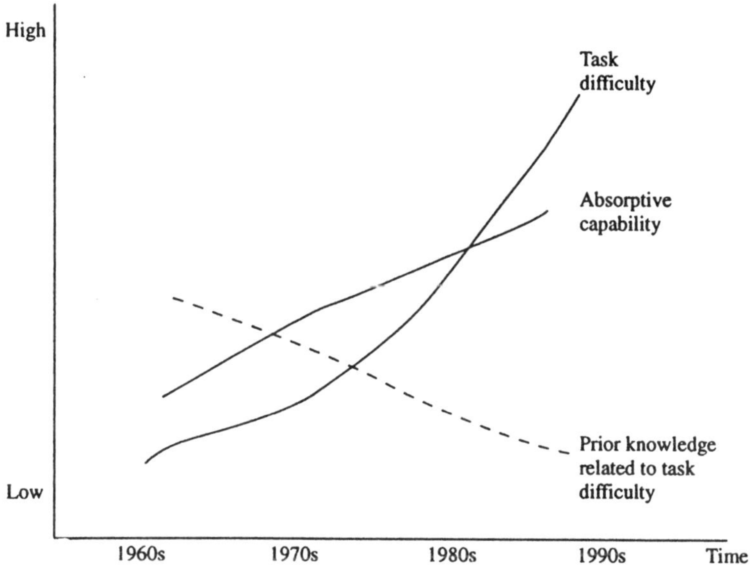

Figure 13.3 Absorptive capacity and task difficulty

face of increasing difficulty in identifying what is relevant, who knows what, who can help with what problems and who can exploit new information, several Korean firms, particularly major chaebol, have set up outposts in Silicon Valley in California to 'leapfrog' into state-of-the-art technologies by monitoring technological changes and to acquire advanced semiconductor and computer technologies. These outposts in California also serve as training posts for scientists and engineers from R&D centres and manufacturing plants in Korea (Kim, 1993).

California provides another advantage to Korean firms by solving the most obvious obstacle to Korea's high technology adventures: lack of experienced scientists and engineers for high technology industries. California is populated by thousands of top-notch scientists and engineers who are Korean-Americans. Leading chaebol have lured away some of the best. Many of them left Korea more than a decade ago, earned PhDs in America's best universities and rose through the ranks of such leading US concerns as IBM, Fairchild, Intel and National Semiconductor. The well-financed Korean chaebol have given them challenging jobs and attractive compensation packages with considerable independence. They serve as a major source of new knowledge and play a pivotal role in developing high technology products.

However, research activities in universities, as mentioned above, have been relatively underdeveloped. For example, universities account for only 5.4 percent of the nation's total R&D expenditures in 1971 and 10.5 percent in 1987 compared with 33.1 percent of the nation's R&D manpower and 78.4 percent of its PhD-level R&D manpower in 1987. The efficacy of a recent move to strengthen universities' basic research capability will remain to be seen. In light of the absence of R&D capability in universities, public R&D institutes have been a backbone of advanced R&D in Korea. They have also been major instruments for national R&D projects, as they are recipients of over 90 percent of research grants awarded by the government in 'new' technology areas.

Fifth, as learning became more difficult, firms started investing in R&D in order to give rise to absorptive capability and to generate new products and processes. Many skills and activities required in reverse engineering have easily been transformed into R&D activities, as they are similar each other. Table 13.2 shows Korea's R&D investment. The private sector has gradually assumed a bigger role in the country's R&D efforts in response partly to increasing international competition and partly to a policy environment conducive to private R&D activities. For example, the private sector accounted for only 32 percent of the nation's total R&D expenditure in 1971 but for 84 percent in 1990. Spurred by various public incentive programs, the number of corporate R&D laboratories increased from one in 1970 to 1201 in 1990, reflecting the seriousness with which Korean firms are investing in 'learning by R&D'. Although small and medium-sized firms account for over 50 percent of the total number of corporate R&D centres, chaebol dominate R&D activities.

Sixth, large-scale, hierarchical, efficiency-oriented, implementation-viable organizations that helped give rise to absorptive capability through 'learning by doing' in the 1960s and 1970s are now paradoxically a major hindrance to raising absorptive capability through R&D in Korean firms. Large-sized organizations must inevitably adopt a bureaucratic system in order to maintain order within organizations but then become slow in decision-making and develop difficulty adapting to a dynamically changing environment. Hierarchical, efficiency-oriented, implementation-viable organizational pattern is compatible with 'learning by doing' but a major hindrance to innovation. Korean firms have to transform existing organizations into a lateral, innovation-oriented, initiation-viable organizational pattern (Zaltman, Duncan and Holbek, 1973; Kim, 1976, 1980b). This is one of the most difficult problems Korean firms have. New recruits from advanced countries are a major source of new knowledge, as mentioned above, but most Korean firms failed to provide a conducive

Table 13.2 Major R&D indicators in Korea (unit: billion won)

	1971	*1976*	*1981*	*1987*	*1990*
R&D expenditures:	10.67	60.90	293.13	1,878.0	3,210.5
from government sources	7.29	39.18	121.73	383.0	511.6
from private sources	3.38	21.72	171.40	1,495.0	2,698.9
Govt vs private	68 : 32	64 : 36	42 : 58	20 : 80	16 : 84
Manufacturing	na	16.70	111.77	1,114.7	2,134.7
R&D/sales	na	0.36	0.67	1.83	1.96
Gross National Product	3,407	13,913	45,528	106,024	206,207
R&D/GNP	0.31	0.44	0.64	1.77	1.91
Number of Researchers (Total)*	5,320	11,661	20,718	52,783	70,503
Govt/Public Inst.	2,477	3,592	5,065	9,184	10,434
Universities	1,918	4,811	8,488	17,495	21,332
Private sector	925	3,258	7,165	26,104	38,737
R&D exp/researcher (W. 1000)	4,306	5,223	14,149	35,580	45,537
Researcher/1000 population	0.08	0.33	0.54	1.27	1.64
Number of corporate R&D laboratories	1	12	65	604	1,201

*The figures do not include research assistants, technicians and other supporting personnel.
SOURCE MOST, *Science and Technology Annals* (various years).

environment for their creative activities. Many chaebol realized the problems of their organization and managerial style and are in the process of making major changes. They find, however, that while organizational structure can be changed overnight, changing the behavior of managers and organizational members to be compatible with a new structure takes a long time to accomplish since these persons have never experienced any other types of organizational or managerial configurations. To introduce a new managerial climate, many Korean firms are now diagnosing and reshaping their organizational culture from a bureaucratic culture to an organic culture.

In conclusion, Korea had enjoyed a position of high absorptive capability (Cell 1 in Figure 13.4) in the 1960s and 1970s. Korean firms had high prior knowledge related to task difficulty and a high intensity of effort. But it is suspected that Korea has gradually slid down towards a position of low absorptive capability (Cell 4) in the 1980s and 1990s. Prior knowledge

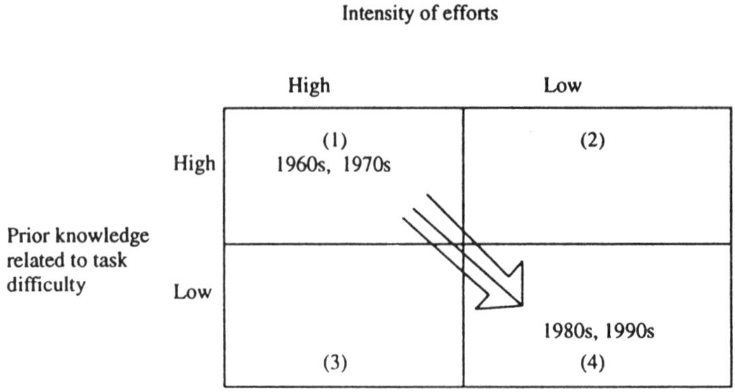

Figure 13.4 Path of absorptive capability

related to task difficulty has declined recently and entrepreneurial and hardworking traits have noticeably disappeared in Korean firms.

What should the government and private sector do to sustain Korea's economic growth in the future? There may be many things that will affect the growth of the Korean economy in the future and both the government and the private sector should take measures that will give rise to relevant prior knowledge and intensify individual and organizational efforts. To mention a few of these measures:

First, it is the government's responsibility, particularly in Korea, to supply highly trained human resources with relevant prior knowledge. The first step for the government is to transform a selected number of leading universities from undergraduate teaching-oriented programs to graduate research-oriented programs. Investment across the board will not be effective, for not all universities should be research-oriented. The selected universities should be internationally competitive in advanced research if Korea seeks to be competitive in high technology areas. The current policy of using public research institutes as a backbone of basic and applied research is not a good strategy, as its impact is less extensive and long-term than that of universities. The research-oriented university generates not only new knowledge but also highly trained human resources with relevant prior knowledge for R&D. These may be channelled into corporate R&D laboratories. Research-oriented universities also produce invaluably important technical entrepreneurs, prepared to establish and manage technology-based small firms.

Second, the government should revise its corporate tax law in such a way that corporate funds can be channelled into these universities. That almost all of the yearly expenditure in leading private universities is met by tuition (which is fixed by the government) indicates how difficult it is for these universities to allocate resources for R&D.

Third, the government should create a conducive macroeconomic environment, in which private firms could strengthen their absorptive capability. The Korean government has already introduced various measures for this purpose, namely, procurement programs for new products, R&D subsidies for risky ventures, joint R&D programs between public and private R&D institutes, tax incentives and preferential financing for knowledge generation and human resource development, etc. What is important is to evaluate frequently the effectiveness of these policy instruments, for policy implementation in a bureaucratic setting can be problematic.

Fourth, the government should take strong measures to foster the emergence and growth of small and medium-sized enterprises, particularly technology-based small firms. Chaebol may still be in an advantageous position in the future in developing such industries as semiconductors and aircraft which require a large size investment. They will, however, be slow and inadaptive to highly dynamic application areas (e.g., microcomputers, software, etc.), where technology-based small firms will have a definitive advantage. Studies show that small technology-based firms were responsible for important innovations, creating a disproportionate number of new jobs (Kim, 1988b). These firms are staffed and managed by well-trained scientists and engineers who are highly motivated to win in competition. Both relevant prior knowledge and intensity of effort are present in these companies. They are also so small that their organizations become organic and free from bureaucracy and labor unions.

Fifth, large chaebol should decentralize their organization, making a number of small, autonomous business units, in order to respond quickly to a dynamically changing environment. Large organizations that have been instrumental in bringing about scale operations in labor-intensive stable industries, are now a major hindrance in adapting to dynamically changing high technology industries. The larger the organizations, the more inevitable is bureaucratization in order to maintain order. In contrast, the smaller the organization, the easier it is to make them organic and adaptive.

Sixth, firms should transform existing hierarchical, efficiency-oriented, implementation-viable organizational structures, which were compatible with 'learning by doing', into lateral, innovation-oriented, initiation-viable structures that would help 'learning by research'. Galbraith (1982) offers several suggestions that will help in the reorganization of existing firms into innovation-oriented structures.

Seventh, firms should retrain managers so they are oriented for innovative organizations. Adopting an innovative organizational structure can be done overnight, but the behavior of managers and organizational members can only be changed through training and indoctrination over a long period of time. An innovative organization cannot be effective unless its managers use skills compatible with its structure and its members follow suit. In-house training could play an important role here.

Eighth, firms should create a conducive climate within large organizations for 'intrapreneurs' to be fostered. Intrapreneurs are highly motivated individuals who exploit their ideas and external information for innovation. Intrapreneurship system may be one of the most effective ways for large organizations to be innovative.

IV IMPLICATIONS FOR OTHER DEVELOPING COUNTRIES

Several things that stand out in Korea's industrialization process offer useful implications for other developing countries.

First, what distinguishes Korea from other developing countries is the way it has invested in human resources even before launching a drive to develop its economy. Had it not been for the formation of trained human resources in advance, Korea's economic development in the 1960s and 1970s would have been much retarded. The Korean government's failure in developing highly trained human resources is now a major bottleneck in taking on the challenge of high technology industries in the 1990s. What this implies is that investment in human resource development should precede industrialization, as human resources cannot be trained overnight when needed.

Second, what is interesting in Korea is that compared with some of the other developing countries, Korea relied least on direct foreign investment and licensing but most on imports of capital goods and informal transfer, such as reverse engineering. Such a policy was efficacious because Korea had developed human capital that could assimilate technologies embodied in physical capital. It also enabled Korea to keep itself independent from the control of multi-nationals. There are three implications that can be inferred. One, developing countries should view technology transfer as a catalytic source of knowledge transfer to strengthen absorptive capability which, in turn, leads to international competitiveness and economic growth. Thus, technological transfer need not be a cause of foreign dependence (Kim, 1991). However, heavy reliance on direct foreign investment as a means of technology transfer may to some extent lead to foreign dependency, especially when local firms do not have absorptive capability. Direct foreign investment transfers knowledge and skills related to production but hardly those related to engineering and innovation, particularly when the parent company invests to exploit the local market in developing countries. Two, Korea's experience indicates that the majority of important and crucial information needed to solve technical problems in developing countries can be obtained, free of charge, through informal mechanisms, if developing countries have local absorptive capability to undertake reverse engineering tasks (Kim, L. and Kim, 1985). What is important is to have local absorptive capability to be independent from multinationals and to minimize technology transfer costs. Three, technology transfer and local absorptive capability building should be viewed as complementary rather than substitutable; higher capability gives rise to better bargaining power in technology transfer and at the same time to rapid assimilation of imported technology.

Third, although the domestic market had been highly protected in Korea to nurture infant industries, Korea's export-led economy has forced firms to exert their efforts to acquire absorptive capability fast in order to survive in international competition in price and quality. One hypothesis is that inwardly-oriented developing countries may not be able to acquire absorptive capability as fast as outwardly-oriented developing countries, as the former may be under less competitive pressure.

Fourth, the process of acquisition in Korea shows that there is an appropriate sequencing in absorptive capability development with production capability being developed in advance of either engineering or innovation capability (Kim and Lee, 1987). An implication is that policy-makers and managers in developing countries can plan ahead in focusing their efforts at the national as well as corporate level on different capabilities as industrialization progresses. Current economic problems in Korea stem largely from its unpreparedness in engineering and innovation capabilities.

Fifth, Korea's experience also implies that sociocultural factors such as entrepreneurship and achievement-oriented traits may not endure for long. Drastic shifts in economic, social and political environments can change such sociocultural characteristics. This fact has both positive and negative implications. The positive side is that a country or firm can create cultural traits such as that of industriousness, or working hard, while the negative side is that such desirable cultural traits are difficult to maintain. A systematic diagnosis of social and organization culture and effective programmes to shape and reshape it should be formulated and implemented in order to make organizations effective in building absorptive capability.

References

Baranson, Jack (1973), 'The drive toward technological self-sufficiency in developing countries', paper presented at the Conference on Latin America–United States Economic Interactions: Conflict, Accommodation, and Policies for the Future, University of Texas at Austin, 19 March.

Business Korea (1990), 'Balancing workers' interests against the nations', February, p. 31.

Cohen, Wesley M. and Daniel A. Levinthal (1989), 'Innovation and learning: the two faces of R&D', *Economic Journal*, **99**, 569–96.

Cohen, Wesley M. and Daniel A. Levinthal (1990), 'Absorptive capacity: a new perspective on learning and innovation', *Administrative Science Quarterly*, **35**, 128–52.

Dahlman, Carl J. and Claudio R. Frischtak (1993), 'National Systems Supporting Technical Advance in Industry: the Brazilian Experience', in Richard R. Nelson (ed.), *National Innovation Systems: A Comparative Analysis*, New York: Oxford University Press.

Duncan, Robert B. and Andrew Weiss (1978), 'Organizational Learning: Implications for Organizational Design', in B. Staw (ed.), *Research in Organizational Behavior*, **1**, 75–123.

Galbraith, Jay (1982), 'Designing the innovative organization', *Organizational Dynamics*, Winter, 5–25.

Harbison, Frederick H. and Charles S. Myers (1964), *Education, Manpower, and Economic Growth*, New York: McGraw-Hill.

Harlow, H. F. (1959), 'Learning Set and Error Factor Theory', in S. Koch (ed.), *Psychology: A Study of Science*, New York: McGraw-Hill, pp. 492–537.

Hedberg, Bo (1981), 'How Organizations Learn and Unlearn', in P.C. Nystrom and W.H. Starbuck (eds), *Handbook of Organizational Design*, New York: Oxford University Press, pp. 3–26.

Kang, Youngkol (1990), 'The rise of Korean Chaebols from the perspective of organization theory', unpublished doctoral dissertation, University of Arizona.

Katz, Jorge and Nestor A. Bercovich (1993), 'National Systems of Innovation Supporting Technical Advance in Industry: the Case of Argentina', in Richard R. Nelson (ed.), *National Innovation Systems: A Comparative Analysis*, New York: Oxford University Press, pp. 451–75.

KEB (Korean Exchange Bank) (1987), 'Direct foreign investment in Korea' *Monthly Review*, October, 3–13.

Kim, Kwang Suk and Michael Roemer (1979), *Growth and Structural Transformation*, Cambridge, MA: Council on East Asian Studies, Harvard University.

Kim, Linsu (1976), 'Technological innovation, environment, structure, and policy implications: contingency approaches', unpublished doctoral dissertation, Indiana University.

Kim, Linsu (1980a), 'Stages of development of industrial technology in a LDC: a model', *Research Policy*, **9**, 254–77.

Kim, Linsu (1980b), 'Organizational innovation and structure', *Journal of Business Research*, **8**, 225–45.

Kim, Linsu (1988a), 'Korea's acquisition of technological capability for internationalization: macro and micro factors', *Business Review*, **22**, 183–97.

Kim, Linsu (1988b), 'Entrepreneurship and innovation in a rapidly developing country', *Journal of Development Planning*, **18**, 183–94.

Kim, Linsu (1991), 'Pros and Cons of International Technology Transfer: A Developing Country View', in Tamir Agmon and Mary Ann von Glinow (eds), *Technology Transfer in International Business*, New York: Oxford University Press, pp. 223–39.

Kim, Linsu (1993), 'National System of Industrial Innovation: Dynamics of Capability Building in Korea', in Richard Nelson (ed.), *National Innovation Systems: A Comparative Analysis*, New York: Oxford University Press, pp. 357–83.

Kim, Linsu and James M. Utterback (1983), 'The evolution of organizational technology and structure in a developing country', *Management Science*, **29**, 1185–97.

Kim, Linsu and Youngbae Kim (1985), 'Innovation in a newly industrializing country: a multiple discriminant analysis', *Management Science*, **31**, 312–22.

Kim, Linsu and Hosun Lee (1987), 'Patterns of technological change in a rapidly developing country: a synthesis', *Technovation*, **6**, 261–76.

Kim, Seok Ki (1987), 'Business concentration and government policy: a study of the phenomenon of business groups in Korea, 1945–1985', unpublished doctoral dissertation, Harvard University.

Mason, Edward S., Dwight H. Perkins, Kwang Suk Kim, David C. Cole and Mahn Je Kim (1980), *The Economic and Social Modernization of the Republic of Korea*, Cambridge, MA: Council on East Asian Studies, Harvard University.

MOST (Ministry of Science and Technology) (various years), *Science and Technology Annals*, Seoul: MOST.

Napier, R.W. and M.K. Gershenfeld (1985), *Groups: Theory and Experience*, 3rd edn, Boston: Houghton Mifflin.

NAS (National Academy of Sciences) (1973), *U.S. International Firms and R,D&E in Developing Countries*, Washington, DC: National Academy of Sciences.

Nelson, Richard and Sydney G. Winter (1982), *An Evolutionary Theory of Economic Growth*, Cambridge, MA: Harvard University Press.

Ozawa, Terutomo (1974), *Japan's Technological Challenge to the West, 1950–1974: Motivation and Accomplishment*, Cambridge, MA: MIT Press.

Porter, Michael E. (1990), *The Competitive Advantage of Nations*, New York: The Free Press.

Schein, Edgar (1985), *Organizational Culture and Leadership*, San Francisco: Jossy-Bass Publishers.

Shishido, T. (1972), 'Japanese policies for science and technology', mimeo, Tokyo: Nikko Research Center.

Shrivastava, Paul (1983), 'A typology of organizational learning systems', *Journal of Management Studies*, **20**, 7–28.

Simon, Herbert A. (1985), 'What We Know about the Creative Process', in R.L. Kuhn (ed.), *Frontiers in Creativity and Innovative Management*, Cambridge, MA: Ballinger, 3–20.

Stiglitz, Joseph E. (1991), 'Social absorptive capability and innovation', paper presented at Korea Development Institute's Twentieth Anniversary Symposium in Seoul, Korea on 1–3 July.

von Hippel, Eric (1988), *The Sources of Innovation*, New York: Oxford University Press.

Westphal, Larry E., Linsu Kim and Carl Dahlman (1985), 'Reflections on the Republic of Korea's Acquisition of Technological Capability', in Nathan Rosenberg and Claudio Frischtak (eds), *International Technology Transfer: Concepts, Measures, and Comparisons*, New York: Praeger, pp. 167–221.

Yakushiji, Taijo (1977), 'Dynamics of policy interventions: government and automobile industry in Japan, 1900–1960', unpublished PhD dissertation at MIT, Cambridge, MA.

Zaltman, Gerald, Robert B. Duncan and John Holbek (1973), *Innovations and Organizations*, New York: Wiley.

Name index